TRASH TALK

An inspirational guide to saving time and money through better waste and resource management

Dave and Lillian Brummet

PublishAmerica
Baltimore

First printing

ISBN: 1-4137-2518-X
PUBLISHED BY PUBLISHAMERICA, LLLP
www.publishamerica.com
Baltimore

Printed in the United States of America

This book is dedicated to the memory of Marlene Brummet. Her whole life, her whole being was dedicated to caring for her family. No other woman could have done better.

ACKNOWLEDGMENTS

Many thanks to:
Brian McAndrew for his ongoing technical and creative support that helped to make this book possible. Vermiculturist Eva Anthony, for sharing her knowledge of worms. Frank and Marlene Brummet, for all the outdoor experiences, positive influences and the continued moral support that has brought us here. Al and Joanne Siewert, who constantly amaze us with their creative ways of reusing items to make a practical tool or device. Thanks to "Beyond Graphix" (www.openminder.com) for designing the book cover. To our many friends and family who have shown enduring support and shared their reusing ideas with us.

PREFACE

"The strength of a nation is derived from the integrity of it's homes" - *Confucius*

North Americans account for only 8% of the world's population, yet we produce 50% of the world's garbage, and consume more than 33% of it's resources. Production processes in our society result in 94% of the materials extracted turned into waste before we even see the product. If everyone consumed like the average North American, we would require three Earths!

Constantly bombarded with negative information about the environment, finances and natural resources can leave us feeling powerless. Not everyone can afford to donate cash or time to a cause. We can be overwhelmed by the immense environmental problems our world faces, yet each and every one of us can do something to help our world—starting right where we are, right now.

Dr. Jane Goodall's E-newsletter reports that 66% of Americans polled would do more to conserve energy and protect the environment if they knew it had a measurable impact. Psychologists have long known that simply performing one small step will aid in defining a positive outlook on life, and will inspire further participation from the individual.

With all this in mind, Trash Talk began in October of 1999 as a series of articles developed to aid in the Zero Waste initiative. By educating through better resource management these articles focused on ways the average family can make small alterations in their lives to affect positive changes for the Earth. Well received in several Canadian publications, the project inspired us to continue our research and compile the Trash Talk articles into a book. The majority of the ideas presented in the book have been put to use in our own home or in the homes of our families, friends, and neighbors. Most of the ideas are relatively simple and do not require any unusual tools or skills.

INTRODUCTION

Zero waste is not just about recycling and the 3 R's. It embraces waste as a resource that creates jobs and new products. Increasingly, more North American cities are taking on the Zero Waste initiative. Many of these efforts are focused on composting, worm bins, and promoting recycling by having more depots available and providing residential blue-box services.

The three R's of recycling (Reuse, Reduce, and Recycle) are often mentioned, but the order in which they are implemented is not often discussed. For instance, even before recycling, a plastic peanut butter container can be reused for various storage means—numerous times, possibly indefinitely. When shopping, buy in larger containers to reduce packaging. There is another, rather unknown fourth "R" to consider. Refuse to buy the brand that has a container that can not be reused or recycled, or that has unnecessary packaging.

In the case of the peanut butter container, it not only serves as a (otherwise costly) storage container, it also stays out of the recycling loop, saving more time and resources that would have gone into reprocessing the plastic. Eventually it might get broken and end up being recycled anyway, but in the meantime, the accumulated savings and benefits are undeniable.

Trash Talk is about implementing the *Refuse, Reduce, Reuse* tactics *first*—in that order—before even considering recycling. In this, we are not trying to pronounce that recycling is a negative approach by any means. Recycling can reduce overall solid waste by 49%, reduce energy consumption by 43%, cut greenhouse gas emissions by 70%, and slash the emission of hazardous air pollutants by 90%. The Environmental Protection Agency did a study of 10

American states that found recycling materials contributed over $7.2 billion, creating 103,000 jobs. In fact, recycling creates 5 to 60 times the amount of jobs than from landfill or incineration processes. By extending the value of our resources, we are saving money for ourselves as well as our community.

Currently only 3.5% plastic, 34% paper, 22% glass and 30% metals produced are recycled. Why then are there not more things being recycled, when North Americans have embraced recycling like no other sport or lifestyle? For one thing, municipal waste is only 2% of all waste generated, and only 30% of municipal waste is actually recovered.

It seems there are many federal subsidies for resource extraction and waste disposal, which directly undermines the market for recycling. Some North American industries that compete directly with recycling, for example, receive federal tax and spending subsidies! The government, no doubt, looks at the subsidies as a business venture and sees possibilities of financial return today— rather than thinking about the financial and ecological gain recycling would bring tomorrow. We need to raise our collective voices and let the appropriate politicians know we want our money re-routed. More importantly, we need to start right where we are—at home. By taking matters into our own hands, we can reduce our own household costs, ease the burden that is upon the recycling industry, and preserve our resources.

REFUSE

Recognizing the power of the consumer's voice is the first step to restructuring your shopping habits. When purchasing, avoid disposable over packaged or individually packaged products, especially those that contain polystyrene foam. Reconsider when purchasing products made from virgin materials as opposed to recycled. You are sending a powerful message to manufacturers when their sales figures decrease.

Refusal to accept manufacturers' standards is very important. For instance: the efficiency of older 2-stroke engines (personal water crafts, all terrain vehicles, and snowmobiles; among many) is very poor. This does not mean an eradication of the product, but simply a refusal to accept their standards. Newer 4-stroke engines are far more efficient, and clean burning with no lack of performance. Consider the automobile industry that has slowly released green energy (hydrogen fuel cell, solar or electric power) and hybrid cars (gas/electric) to their production lines in the last few years. The trend has really bloomed and we can expect to see many more improved variations available in the near future. You, the consumer, helped accomplished that.

Ensure that you are heard by writing your concerns or ideas to the editors of various publications, the Minister of Environment, and to manufacturers. You may scoff at this, but one letter campaign we took part in resulted in over 10,000 Canadian participants—numbers like that can not be ignored. Let the powers that be know you want reduced packaging that is designed for recycling. Ask for clean energy policies that lean toward reducing pollution. Remember, what all it boils down to is this: they work for you.

Purchasing Eco-certified products and bulk amounts are other sound environmental choices. Did you know that the average grocery store fruit or vegetable travels 800 miles before you purchase it? Frequenting local organic farms helps reduce this avoidable use of energy.

Keeping an eye on the second hand stores can actually make good *cents*. Older tools are often of higher quality than many newer ones now available in stores, so it is worth attaching a new handle to an old head. Bargain prices on near new clothes can be found at second hand clothing stores, many of which donate their proceeds to a worthy cause. Refusing to buy cheaper items that are not durable to the test of time will ultimately extend your shopping dollar and the life of the landfill.

REDUCE

Start by reducing the amount of waste packaging that comes with consumer goods. Choosing to buy large sizes of concentrated products is one consideration. By purchasing rolls of photo film in 36 rather than 12 exposures, you will reduce packaging alone by 66%. Strive to find goods that come in sturdy reusable containers. By buying in bulk at the grocery store, you avoid any packaging other than the plastic bag, which is reusable. All of these actions convince manufacturers to meet the consumer demands. They seem to hear money quite clearly!

Reducing the waste that is destined for the landfill begins simply by asking a few questions before discarding an item and purchasing another. *Can I donate this or reuse it in any way? Can it be repaired, recycled, or composted?*

Extending the life of current landfill sites ultimately reduces the need to create new ones, and helps to keep municipal waste collection costs down. The fact that a reduction in consumption lowers costs is not lost on the business world. Bell South Telecommunications printed its customer bills double sided, reducing paper use by 1.2 million pounds and cutting postage costs by over $530,000. Xerox introduced reusable containers and pallets, saving $15 million and 10,000 tons of waste annually.

As is apparent, the word *reduce* has more than one meaning when discussing proper waste management.

REUSE

What most of us may not realize is that reusing can save a lot of money. For instance: a compost system converts organic waste into a very rich fertilizer, which is reused to grow healthy floral and food crops. Salvaged lumber from a demolition project may often be obtained for next to nothing. Turn a wine box bag into a backpacker's collapsible canteen. And, of course, the now famous peanut butter container stores bulk foods in the kitchen, craft supplies in the kid's room, or nails in the workshop.

Reusing can involve many activities. For example, thoughtful people may volunteer their time to collect, repair, and re-distribute items. We know of many individuals who refurbish and donate toys, cloths, and bicycles to those who need them the most. To reuse can also translate to:

Buying or selling in the used market place.
Renting, lending, donating or borrowing items.
Visiting Reuse sites.
Finding creative uses for waste items.

Reusing will decrease the volume of waste contribution, prolonging the life of the landfills. A good example of this is in Santa Monica, CA where their reuse facility at the landfill reduced the volume of waste by 20%.

Our financial consultant tells us that if people could find a way to save just $7 a day, they could afford to contribute monthly to an RRSP. In an age where financial security is uncertain, where the government continues to gouge us with taxes, and budgets are becoming tighter and tighter, reusing is starting to make a lot of *cents*! If one looks at these actions like coupons—sure it is 50 cents here, 25 cents there, but at the end of the week you might save ten or twenty dollars, maybe even more.

Although we discuss many other waste and consumerism reduction methods, *Trash Talk* is essentially about reusing items formally destined for the landfill or recycling depot. It is not about a quick fix solution; it is about changing consumer's mind-sets by providing ideas that inspire participation from the ground level. There are literally hundreds of ideas we will be sharing with you in this book. We hope not to overwhelm you, but instead to provide encouragement by showing you the direct affect these actions will bring. We have also provided, in the final chapter, resources for you to continue the journey to better waste management. Do not expect yourself to be able to incorporate every idea mentioned at first. The thing is to start by trying just one thing at a time. Then later, try another. Just start where you are. It is how we began, too. We hope you have fun learning, experimenting and hopefully—influencing someone else!

TABLE OF CONTENTS

PART ONE

BAGS

Production of plastic grocery bags actually consumes 40% less energy, generates 80% less solid waste, produces 70% less air pollution, and releases 94% fewer waterborne wastes than paper bags! Both can be reused and recycled, but paper bags can also be composted.

Plastic bags accumulate quickly and many people want to reuse them, but they are soon bulging uncontrolled from our cupboards and drawers. There are now storage containers available to organize this, to be either mounted to a wall, or hung from a hook. They have a hole at the top where bags are contributed, and a hole at the bottom where they can be pulled out. We use two of them in our home, one for grocery bags and one for bread bags.

Grocery Bags

Many people bring their own cloth or canvas bags for their grocery shopping. Gudrun, our friend from Germany tells us there is often a deposit on shopping bags, and so the use of cloth bags is much more common throughout Europe. Relatively inexpensive, cloth bags can be washed and reused many more times than plastic and are unlikely to tear when full of groceries. The stores will have an easier time maintaining prices if there is a reduction in plastic bag demand. Cloth bags are an excellent advertisement feature and would make fine customer appreciation gifts for your business with your logo on them.

The plastic grocery bags have so many uses in our home that we often run short of them. The most common reuse is for wastebasket liners. Why buy wastebasket bags for dry waste areas when we get them for free? A box of 24 bags costs $3.50 (more when tax, shopping time, and fuel spent getting to the store are figured in), and the average house consumes 2-3 boxes a year. So just by lining your dry wastebaskets with grocery bags, you could save $10.50 a year. It does not sound like much, but this is just one example of the many uses for these bags.

We use them to carry various things during transport for items such as used clothing that we are donating, or books we are bringing to a friend. We take them to farmer's markets when buying fresh produce. Use them to store your own garden harvests in the fridge, or to send off with visitors. We take them to dog-friendly trails to donate to the bag dispenser at the trail head. Hopefully it may encourage other owners to clean up after their loved ones. For those of you who have extras, you may want to consider donating to the Food-bank or thrift stores, which are often in need of bags.

Use when camping to store laundry and to keep items dry. Because they take up very little room crumpled into tiny, weightless bundles, we pack them inside paper towel and toilet paper tubes. To reduce weight in the backpack and prevent tempting scavengers, we make our camping and trail meals from dry ingredients. Dried meals only take a few minutes to absorb the hot water—reducing fuel consumption, lingering odors, dishes, and cooking time. We often cut bread bags to fit the size of the dried meals or snacks and then double bag the lot of them as an added insurance. This storage method reduces food odors, thereby decreasing the chances of enticing wildlife to the bear bag or backpack. On the return hike, carry a bag in hand and scan the trail for garbage. – see *Clean Walking.*

Years ago, we watched a news special on an elderly lady who made grocery bags into rugs. She tore the bags into rough, uneven strips and crocheted the strips together. She began the hobby just for herself, but the rugs became so sought after that she was kept busy with a constant supply of bags and requests. Looking at the rugs you would have never guessed they were made from recycled bags. Ruth, a very creative friend, used her over-abundance of bags to stuff her valance curtains.

Clear (Bread) Bags

Smaller plastic food bags usually come clear or translucent and, unlike shopping bags, have no holes. Sealed plastic bags (from frozen peas or powdered milk) can be cut open and reused many times as well. However, thin produce department bags often use water-soluble inks, so we do not wash them or reuse for food.

Using a sink of fresh hot soapy water, open the bag and swish around. Fill it about 3/4 full with the soapy water, grip the top together, lift above the sink and move it about to check for leaks. Where there are holes, there will be a steady stream of water. Throw those bags out or reuse them where a perfect seal is not required.

Rinse the bags and hang to dry. You can do this a few different ways. The simplest is to droop the bags over a full dish rack allowing them to drip dry. Or, pinch one corner in a cupboard door over the sink or dish rack. Some people have strung lines for the bags to hang with a clothespin but, again, it should be over a sink or drip tray. If you have the time, they can be dried by hand but a little moisture will remain, so leave the bags out in the open air for a while. In a few hours, turn them inside out so they dry completely before storing.

We reuse plastic yogurt containers to freeze foods but because they do not have a seal suitable for freezer use, we make sure by doubling up with one of our bags. After placing the full container in the bag, gather the top together and suck the air out before sealing it with a recycled twist tie, or simply tie a knot in the bag. Placing your sticky label on the bag eliminates the problem of getting labels off the plastic containers.

Use to store garden produce and home baked goods. At our home, they are used all the time for lunches on the go. Used for pet waste, used oil, meat scraps, and bones, bags will help contain and isolate offensive odors in the garbage can. A smelly can tends to be taken out whether it is full or not, resulting in more garbage bags being used.

Clear plastic bags can also be utilized in the greenhouse. This is where those leaky bags come in handy. We cover flats of newly sown seeds with a bag that has been cut open so that it lies flat. This provides a mini-greenhouse effect and keeps the moisture in. As the seedlings grow, prop up with little sticks so that the plastic does not sit on the leaves and cause rot. Insert plant pots into plastic bags to eliminate the need for drip trays. When cloning, create a mini-greenhouse by bringing the bag up over the cutting, and close with a twist tie.

Zipper lock sandwich and freezer bags can be washed and reused more than a dozen times. Let us quickly estimate how much this one simple act saves the household money. At $2 a box, reusing the bags a dozen times saves your home $24. We are avid gardeners and between food storage and lunches, we save about $50 a year by reusing zipper bags. For this reason, we purchase the best quality we can find on the shelf. Recently, snack manufacturers have started to use strong foil bags with a zip top, making them ideal for reuse.

Food bags are not the only kind one can reuse. Our pet food bags are reused for sorting recycling, and to line the larger workshop waste bucket. Bags used to protect mailed magazines work well for containing odors in the garbage bag. The list goes on and on.

Benefits
- Extend the life of the landfill.
- Save money by reducing the number of bags you need to buy (along with their packaging).
- Reduce the number of trips to the grocery store.
- Cloth bags:
    - are very unlikely to tear when full of groceries.
    - can be washed and reused many times.
    - make an excellent advertisement feature.
    - make fine gifts.
- Stores will have an easier time maintaining prices if there is a reduction in plastic bag demand.

BRUSHES AND BROOMS

In 1995, 19 million tons of plastics were sent to landfills—after recycling was employed, that number dropped by 1 million tons.

Dentists recommend that we replace our toothbrush 4 times a year, but that results in over 100 million pounds of toothbrushes contributed to the landfill annually. Many companies are rising up to meet this landfill challenge in different ways.

Toothbrushes containing up to 70% recycled polypropylene and plastic materials are now on the market. Another toothbrush on the market is made from recycled yogurt cups. Some manufacturers have opted to make the handles with replacement heads. Toothbrush subscription services use an innovative approach to the issue. Members receive new toothbrushes made from post-consumer materials by mail. In turn, the members send their used brushes back for recycling into plastic lumber. All of these are excellent consumer choices.

The recycling community is quite excited about plastics. With the present popularity of plastic, the demand for raw materials will only increase and eventually, recycled plastics will serve as a necessary resource. The recycling process does nöt shorten the grains, strands, or fibers within the material and consequently does not reduce its strength. Therefore, there is high potential for recycled plastic products.

Experts predict that in just 20 years, all plastics will be recyclable and we will have viable markets for the recycled material. Such news is encouraging for the environmentally-minded. In the meantime, reuse of toothbrushes is still an option. Lillian ran a residential cleaning business for 6 years that depended on what those in the industry call *The White Glove Treatment*; it means every crack is scrubbed clean and every surface is highly polished. Toothbrushes are wonderful cleaning tools. They work especially well around the tiny crevices of toilet lid hinges, around faucets and drains. As polishing tools, toothbrushes work well for shoes, especially in areas hard to reach. In the workshop, they are the best tools to clean threaded parts.

Brooms can easily have their lives extended in a few ways. Simply washing the broom a few times a year can really extend its life. Yes, we said washing. That means reusing the water left from washing dishes or the water gathered in a bucket while waiting for the right temperature in the shower. Regardless of where you get the water from, it needs to be warm and soapy. Dip the broom in and swish around a lot. Pretend it is in a washing machine. Allow to dry. Keeping the bottom bristles scraped off with the serrated edge located on many dustpans will reduce hair and other debris from accumulating. When brooms become ugly on the bottom, trim the bristles down to where they look clean again with a pair of strong, sharp scissors. This gives the broom a whole new life. We have extended the life of our kitchen broom many times over, by doing this simple task. Before the broom is too short to use any longer, it is time to retire it to the workshop or garage. The shortened bristles are very stiff—great for keeping cement floors, steps, and other rough surfaces swept clean. By the time the broom has worn completely out, it looks haggard and awful. However, there is more reuse ahead for the handles of brooms. Remove the handles and set aside while you read the chapter *Rods and Handles* for some ideas on what can be done with them.

Benefits

• Choosing plastic toothbrushes containing post-consumer materials is a way to support the reuse effort, and reduce plastic

waste.
• Reduce consumerism and save money by reusing old toothbrushes as cleaning tools.
• Extending the life of brooms will save money while reducing waste.

CARPETS

When planning to install new carpets, consider the modular ones that are now available in 18" squares instead of a 12' wide roll. The benefit comes in effect when looking at long-term needs. These tiles allow us to replace only the worn or stained areas, rather than the entire floor space—reducing consumption and costs by 80%.

Used carpets, or remnants of carpets, can be found at most house renovation sites, and a myriad of different shapes and sizes of new carpet sections can be attained for very cheap at carpet installation stores as remnants.

Place a section in the back of the van or in the trunk of the car to protect the interior. Small pieces can be placed under furniture wherever scratching or indenting delicate surfaces is a concern. We like to custom cut each piece so that very little is visible to the casual eye.

Using a piece in front of the cat's litter box will prevent litter from being tracked around the house. Sections placed at high traffic areas such as entrances can greatly reduce the amount of cleaning necessary inside the building.

Many homes have concrete floor areas, such as in the garage or basement. To ease the stress on backs and feet, lay sections of carpet at working stations. Remnants of new carpet can cover the main area of a room effectively for around $40 at a carpet store. Since people do not usually move furniture or walk against walls, only the central floor space really need carpets.

We reuse old carpets outdoors as well. A section placed in front of the tent entrance allows for much more cleanliness and comfort. In the garden, strips of old carpet make effective paths. Lay the pile side down on your garden pathways, so that the backing is the pathway surface. These work extremely well as a weed barrier that also keeps your feet clean.

Benefits
- Modular carpet squares reduce consumption and costs.
- Save money by purchasing remnants at carpet stores.
- Protect the interior of vehicles.
- Prevent scratching or denting to delicate surfaces.
- Decrease the amount of cleaning necessary for high traffic areas.
- Prevent muddy and weed ridden garden paths.

CEREAL

Hectic schedules with early morning activity allows very little time for cooked breakfasts in our society, except maybe on weekends. This results in the incredible number of boxed cereal we consume. Instead of looking at the packaging as waste, try seeing it as a valuable resource.

The boxes are made of paperboard and some reuse ideas are discussed in the chapter of this book entitled *Trees Please.*

We rarely purchase rolls of waxed paper because the liner bags from cereal boxes are reusable. Remove the bag, wash in a hot soapy solution and hang to dry. Whole, the bags could be used for wrapping sandwiches, like our grandparents did before we had so many plastics available. Or, open completely by cutting out the seams. Use these large sheets in place of other wax paper needs in the kitchen, like lining trays when making candies or dipped cookies. Whole sheets can be used to protect working areas when doing any kind of messy craft, such as candle making.

We also cut them into squares roughly 3-4" in size and store in a reused zip bag. These squares are used to separate items like veggie patties or burgers when freezing in zipper bags for future use. Use in a similar manner to separate treats like brownies or other deserts.

The crumbs at the bottom of the cereal box do not have to go into the compost pile. We reuse them as an ingredient in cookies in place of a portion of the seeds, nuts or oats. As an addition to breadcrumbs, cereal can make a flavorful coating or ingredient for veggie burgers and meat loafs. Simply keep a small container in the kitchen cupboard and contribute cereal as necessary—it is surprising how quickly it accumulates—and gets used up again!

Benefits
- Purchase fewer rolls of wax paper.
- Save money by reducing need for sandwich bags.
- Protect craft and work areas.
- Extend the value of the shopping dollar and increase nutrient content.

CLOTH

Reusing fabric is far from a new activity. At one time, Europeans used rags to make writing paper—until there was a rag shortage, that is. Today, many companies produce products made from recovered cloth scraps, but very few (if any) accept used fabric. However, many individuals have quietly taken up the challenge for generations. Historically, gardeners have successfully composted shredded rags in just a few years. We have tried this and have had varying success; it works best with loose weave cloth that is made from natural fibers.

The rising number of people that are choosing to buy used clothing can be seen by the increasing number of consignment and second hand clothing stores. These places are providing an ethical service by creating employment, and often donating funds to various charities. Used clothing outlets also reduce the pressure on landfills and help families to save money. Locally, we have a thrift store where the volunteers repair, wash, and sell the donated clothing—raising funds for the local hospital.

Lillian's mother, Joanne, a single mom with 3 children, learned the value of shopping at used clothing stores. She shopped for large dresses and other such clothes because they provided a wealth of material. After removing the zippers, buttons and seams, she would make new clothes and blankets for her family, and rags for her cleaning company out of the salvaged cloth. Although her own three kids should have kept her in supply, she used to gather old jeans from friends. She even gathered jeans from thrift stores, the ones they could not sell because the zipper did not work or were otherwise beyond repair.

"They were always glad to get rid of them as it kept their disposal

28

costs down," Joanne explains. "I felt good about reducing the landfill while providing things for my family that we could not afford." A *friend* of Joanne has also reused fabric to make children's clothes, aprons, and doll clothes, which she sold at craft fairs.

Our favorite use for jeans with worn knees is to remove the legs and create shorts. We save any good pieces from the legs to use as patches. Tie the long seam section into a knot and use as a very durable pull toy for dogs. All kinds of garments for children, such as skirts, vests, jackets and jumpers can be fashioned from pieces of recycled adult jeans, but they may also require a lining. The pelvic part of jeans, can be sewn (front to back, straight across at the crotch level) into a purse, or a bag. A recycled belt makes a fine shoulder strap. Or, cut out the back area, run an old belt or a strap through the loops and—voila!—a ready-to-use apron with pockets, without sewing one stitch!

Strips of cloth linked together can be crocheted into rugs. The width of the strips will depend on the thickness of the fabric being used. Make balls of like-colored material so that you can design patterns with a particular color. Jean rugs work well for heavy-traffic areas. Cotton and cotton-blends make more delicate rugs that are more appropriate for light traffic areas. Wool and wool-blends work well as living room mats. Nylon stockings can be reused too. Because they are mildew-resistant and dry quickly, they are best for kitchen and bath mats. Also, keep in mind that most fabrics require the cut sides to be folded under to prevent fraying; for nylons that won't unravel and don't need strips cut, just use the whole leg. For more, see *Nylons*.

Pieces of cloth sewn together patchwork-style make great covers for throw pillows. Use large pieces of material to create drawstring bags. Small bags are favored by children who use them to keep their treasures in, but are also useful as potpourri sacks, and seed storage bags. Larger ones can serve as gift bags or stuff sacks—especially handy for those that like to camp or backpack. These bags make excellent containers for first aid supplies, shaving kits, laundry, and more. You can even reuse a zipper to close the bag, rather than a drawstring closing. Dave prefers the strings with a plastic cord lock device, as they do not have to be tied shut, and are easily opened.

Garage sales are great places to find used blankets, which can be useful in so many ways. Even tattered blankets can be reused. Folded in half and sewn together, thin blankets make excellent pet beds or, as is, are often gratefully accepted by many animal shelters and wildlife rehabilitation centers. Kept in the vehicle, a blanket can also be brought out for picnics and emergencies, while traveling. Blankets can be put to use in the garden as extra cold protection on cold frames, row covers and greenhouses. If in good shape, charitable organizations and shelters for the homeless would be grateful for donated blankets and towels.

Lillian ran a cleaning business for 6 years, and she found that flannel sheets and terry cloth towels make the best rags because they are absorbent. She always made sure that the edges of the rag were folded under and sewn securely to prevent unraveling. Those threads can easily catch on items when cleaning. Large towels can be made into face cloths, hand or kitchen towels, or several squares can be sewn together to make a simple potholder.

Clothing should not be wasted and there are many options to get the most out of this resource. Any good clothing can be given to people you know, donated, traded, or sold on consignment. Please, do not just send it to the dump!

For some ideas on using fabric as gift wrapping, see *Earthly Holidays*.

Benefits
- Reduce pressure on your community's landfill.
- Save money by creating new clothing and rags from old material.
- Never buy dishtowels, face cloths, or potholders again.
- Extend the gardening season with old blankets.
- Buy and sell at used clothing outlets to save money.

CONTAINERS

In 1995, 50% of North Americans had access to a plastic recycling program, yet only 20-25% of the material was recovered. Plastic production goes up an average of 10% annually.

Beverage containers account for roughly 5% of the waste in landfills. Using more recyclable and returnable containers would greatly reduce this number. Only a few containers are returnable for deposit, but the trend for charging deposits is now on the rise. With 114 billion beverage containers wasted annually, corporations now agree we have a problem and are willing to work toward a solution.

Nowadays most plastics have a number stamped on them to designate their type for the process of recycling. It is important to note that even a small amount of the wrong type of plastic in a melt could ruin that batch. Until recently, not all plastics were accepted at all depots. However, clear plastics such as ketchup bottles, milk containers, and juice jugs are readily taken at most recycling centers. In our local area, rigid plastics numbered 1-7 are now accepted. Manufacturing plastic containers numbered 3-7 is the most polluting process in the industry; they are also the most difficult to recycle. Therefore, we recommend avoiding these types of plastics when possible.

Containers need to be rinsed clean of food residue before taking in to the depot. Food rinse water is excellent fertilizer for landscaping and compost piles. If you do take advantage of the recycling program—be proud. Because you are doing this simple chore, you have helped extend the life of landfill sites, reducing the need for new ones, and thereby keeping the municipal waste rates from increasing.

Recovered plastic is washed, dried, and ground into flakes. The flakes are used in the manufacturing of drainage pipes, bathroom stalls, kitchen drain boards, combs, fleece fabric, blankets, upholstery, carpet fiber, plastic lumber, stuffing for clothing, furniture, and automotive parts.

While we wait for more containers to become returnable, there are many creative ways to reuse them. For any ideas that involve exposing a cut edge, we recommend that you file any sharp barbs with a few quick strokes of sandpaper and/or tape the edge. This trick makes the edges safer and stronger.

Jugs can be filled with water, juice, or stock, and put in the freezer. No matter which liquid you choose to freeze, allow at least 2" of expansion room. In a possible emergency, they will help to keep your freezer cold, while providing you with good drinking water or nutritious broth as they melt. Frozen water jugs work well for the picnic cooler too. They replace the need for bagged ice, thereby saving money while reducing waste. At as much as $3 per bag of ice, people like us, who do a lot of camping, can save about $50 annually doing this very simple step. Moreover, we get nice cold refreshments as they melt; when camping, something cool to drink may be hard to come by. There is also less mess, as bagged ice usually leaks throughout the cooler.

The small 16oz bottles work great as water bottles when out hiking or cycling. They can even fit in some jacket pockets! Lillian takes one everywhere she goes, even shopping. She finds it is a good way to increase water intake. Bringing along a water bottle eliminates the impulse to purchase liquids while in town, keeping the budget in check. While out walking the dog, this water jug comes in handy for him as well. One of us packs a lid (from a peanut butter jar) in a pocket. This makes a great water dish. It has also helped us teach him not to drink dirty water, which could make him sick.

Make a ground dwelling rodent deterrent by reusing large plastic pop bottles and metal rods or piping. Drive the rod 2-3 feet into the ground. Place over the pipe, an upended pop bottle that has had vanes cut in the sides. The vanes will catch the wind and rattle the bottle against the pipe—emitting a bothersome vibration into the ground that deters moles and voles.

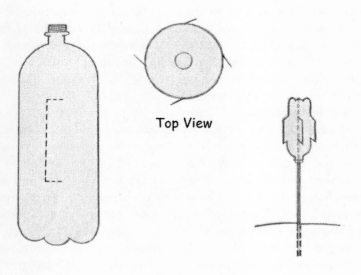

Top View

Jugs can be made into inexpensive flying insect sticky-traps that can be reused over and over again. Simply paint a jug bright yellow, place it inside a recycled plastic bag, and close with a twist tie. Coat the bag in *Tangle Foot* or petroleum jelly and place the homemade yellow sticky trap in amongst affected plants. We have put some soil or sand in the jugs to make them sturdier when placing on the garden bed, but leave them empty if we plan on hanging them up high on a stake or on a plant's cage. When there seems to be a lot of pests stuck to the plastic, simply remove the bag, turn it inside out for easy disposal and start again with a fresh recycled bag. This small step not only reuses items formally destined for the landfill, but also eliminates the need for purchasing sticky traps. Painted yellow or orange, the lids from plastic food containers can also be reused as insect traps. Drill a small hole in the lid and attach string to hang near the plant. Cover with *Tangle Foot* or petroleum jelly, as before, and hang in the plant. One wise elderly neighbor covered his lids in plastic wrap before coating, allowing him to reuse the lid again.

To make a very sturdy door stopper out of a recycled jug, simply fill with sand. If using clear jugs, you may want to fill them with different layers of colored sand. On the other hand, you may want to get creative and paint them, or coat them with a modge-podge design. Children will love this sort of craft activity.

Try this water saving technique: by placing one or two 16oz bottles of water inside the toilet tank clear of the inner workings, it will displace the water level in the tank, reducing water use per flush. For more details see *Water Use*.

Plastic peanut butter or mayonnaise jars are large and bulky enough to quickly fill your bag for curbside pickup. They are marked recyclable, but they may not be accepted at all recycling centers. In our house, we never throw them out - we need them too much.

Around the shop, rigid plastic jars are indispensable for storing razor knives, nuts and bolts, and other metal parts. We also use large jars for the storage of garden seeds and soil supplements.

In the kitchen, larger jars are reused mostly for storing bulk foods and our own dehydrated goods. Cotton balls, detergents, and a host of other things around the home can also be stored in these jars.

One-gallon plastic milk jugs are readily accepted for recycling but there are many uses for them even before they hit the blue box. In the garden, they can become mini-greenhouses. Simply cut off the bottom of the jug (the pieces can still be put in the recycling box) and take the lid off for a little vent at the top. Place the jug over the transplant and push the cut bottom edges into the soil. Heap up a bit of soil on the outside for additional support. If the jug-house becomes too warm for the plant, either remove it during the heat of the day or paint with white wash to reduce the strength of the sun; or cut an extra venting hole in the side of the jug, eliminating the need to hover over the plant, making sure it does not get overheated. This is usually unnecessary for translucent jugs. If you find it is not sturdy enough for the winds in your area, use a few stakes to hold in place.

An excellent way to fertilize vine-plants is achieved by cutting the bottom of the jug off, then burying the jug spout down into the soil beside the plant you wish to feed. When the jug is in place, fill with 2" of sand or fine gravel. Then fill the jug with humus (composted matter), or worm castings. Fill the container with water twice a week, letting the nutrients seep down to the roots of the plant.

Use as a deterrent for deer and other skittish creatures by stringing them along fences by their handles, loosely enough that they will bang against each other in a slight breeze. Be sure to remove the lids so that the wind will whistle through them. It is a good idea to poke a few holes in each bottom so that they will not fill with rain. This is most useful for rural homes where the view is undisturbed by the jugs.

Paint jugs black or dark red, fill up to 2/3 with water and place in the greenhouse and under the row covers in the garden. The water heated by the sun during the day will retain that heat and slowly release it at night. You can eliminate any worry about paint chips tainting the garden soil, simply by inserting the jug in a clear plastic bag.

Jugs can also be made into plant pots. The top should be removed down to the handle with a razor knife. Again, recycle the debris if possible. Poke a few drainage holes in the bottom. Fill up to 3" with fine gravel, sand, or Styrofoam chips for proper drainage. For a drip tray, reuse a sheet of foil to wrap the base of the jug. A plastic lid or a tin pie plate will work just as well; the bottom cut off a jug from the mini greenhouse idea can be made into a drip tray for this idea.

Milk Jugs and other jugs with handles actually make good harvest buckets. Cut an opening in the top corner so that the bottom, sides, and the handle remain. Loosely hang the jug by a belt at your waist and when it fills up, it is easily tilted to empty it into a larger container. We love these harvest buckets because it allows us the freedom to use both hands.

35

This bucket could be used to store nails or bolts, or anything in the workshop. Lillian particularly likes to hang the bucket on our clothesline to store our clothespins. It has a roof, of sorts, to shelter the pins from rain. We also tied one to the inside of the canoe as a bailing bucket.

Rigid plastic jugs, like those that contain cleaning products, make the best scoops—we have found. Hold the jug by the handle, which should be facing up toward you. Now cut the bottom off at an angle, creating a scoop with a longer bottom edge. You can make the scoop any size by varying the cut or choosing from a myriad of bottle shapes and sizes available.

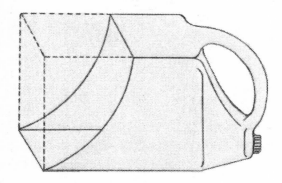

Try reusing yogurt and cottage cheese containers as plastic collars for transplants in the garden. Using a razor knife, carefully cut out the bottom of the container. Place over the transplant in the bed, and press the cut side into the soil firmly. Gently twist the container while pushing in the soil to work it in more securely. This collar will deter mice and other stem pests such as cut worm from reaching your transplant and can be left on all year. Or they work well as plant tags, too. For details, see *Around the Garden.*

Reusing the containers as pet dishes works quite well. They can also be reused as scoops for pet food. The small 8oz containers work well for ant traps. We fill them with borax and white sugar, put the lid on, and cut small holes in the sides for the ants to access the food. They will take this home to their queen.

We find an endless need for storage containers around our house, for everything from bulk foods and crayons, to nails and toys. Food grade containers can store leftovers. We also use them as freezing containers for our soups, stews, sauces, pesto, salsa, tomato pastes, vegetables, and soup stock. When freezing, be sure to leave about an inch of space at the top rim to allow for expansion. Wipe clean, seal the lid, and then place in a recycled bread bag to protect against freezer burn. Suck the air out and close with a twist tie. Use a label, or write the contents and the date it was frozen directly on the bag.

Here are some ideas for labeling the reused containers: Labels can be found at dollar stores for very cheap, and office supply stores sell them in sheets as well. However, unused videotape, cassette tape, and diskette labels are even less expensive, since they are free. Yet, sticky labels are very difficult to remove from containers, and are difficult to wipe clean with a damp cloth. So we prefer using scrap paper and wide packaging tape instead, which can be wiped clean, and is easily removed in hot soapy water.

Our friend, Brian, uses the large square plastic buckets from kitty litter for garbage bins. He cuts out all but the rim of the lid, so when snapped on, the bag is held in place.

We have reused similar buckets around the workshop and yard. Gallon buckets make ideal compost buckets for the kitchen and nearly any container can become a plant pot simply by drilling a few drain holes in the bottom. Small square containers can be used as desk drawer organizers or to store scraps of paper in them for note taking or list making.

No matter what container you are using for boxes, plant pots, wastebaskets and the like—they are all easily decorated, or simply set an undecorated container inside a woven basket instead.

Reuse baby bottles in the garage. They are marked in easy to read increments and are valuable for measuring powders and liquids. Baby wipe containers have a hole in the lid that is perfect for dispensing yarn, and to keep it from tangling. Breath mint containers work well as seasoning containers for picnics and camping.

Deodorant, film and pill containers, and medicine bottles can be reused for many things as well. But it is important to note that these plastics are not food grade. They are handy for many other things, though, such as:

A traveler's sewing kit, first aid kit, or to hold small things, like crayons.
To store tacks or seeds.
Store change.
Mix paints.
Store suntan lotion/hand lotion/soaps or shampoos when camping or traveling.
Use as a cat toy—add some beads or buttons for noise, then glue the top on.

Benefits
• By reusing these plastics that come in all shapes and sizes with tight fitting lids, you can avoid buying a lot of costly storage containers.
• By finding creative uses for plastics formally destined for the landfill, we can lessen the demand on the plastics industry.
• Save money by reducing the consumption of costly plastic products.

DRYER SHEETS

Dryer sheets will not compost and are often just thrown out with the trash. By using only a half sheet at a time, you can greatly reduce your consumption of them. Saving used sheets until a batch of four is collected, it is possible to get one more dryer run out of them. Doing these two small steps, our dryer sheet cost is reduced by nearly 60%. It is worth mentioning that loads containing towels or bathrobes create too much static and require a full, new sheet.

We once heard a reuse suggestion of dusting with spent dryer sheets. Having tried it, we found they are not as effective as a dampened cloth rag. They can, however, be reused in dresser drawers and trashcans around the house where they will release any remaining fragrances.

The exhausted sheets can be reused further, as they do make an excellent fire starter. So does dryer lint, by the way. Both lint and threads, separated by pinching small pieces at a time, can be put in the compost bucket or reused as part of the bedding for a worm bin.

In the last few years, we eliminated the need for dryer sheets all together. Now we use *MelaSoft*—an earth-friendly liquid softener by Melaleuca. It is so highly concentrated that only a tablespoon per load is required. Pour the liquid softener on a rag or sock intended solely for this purpose. Then roll the rag up, giving it a squeeze so that its fibers fully absorb the softener (otherwise run the risk of marking the laundry), and place in the dryer along with the clothes. Now, no dryer sheets are going to the landfill, and at pennies per load, this option will save the household even more money.

Apparently, the need for softeners can be eliminated all together if vinegar is added to the rinse cycle. The trick is to be around to catch the start of the cycle, which is not feasible in most people's situation.

Remember, choosing concentrated laundering products is ecologically responsible because it reduces the waste incurred in packaging, extends the value of your dollar, and reduces the number of purchases you have to make. Choose those with biodegradable, phosphate-free, and naturally derived ingredients, because they do not contribute to the pollution of the environment. With these kinds of mild detergents, a water recycling system is also possible. For more on Grey Water recycling, see *Water Use*.

Benefits
- Reduce dryer sheet consumption by at least 50%.
- Save money and reduce waste by eliminating dryer sheets altogether.
- Concentrated products of naturally derived ingredients decrease packaging waste, extend your shopping dollar, reduce the number of purchases, and will not pollute the environment.

FOR THE BIRDS

Normally, we try our best to attract wild birds to our land, not just because they are beautiful, but because they are terrific at keeping the pest population down, and they also consume copious amounts of weed seeds. For areas with large populations of mosquitoes, black flies, and gnats, attracting birds is a natural solution. It is also important to note that except for a few exceptions, North American bird populations are declining in disturbing numbers every year. Therefore, it is even more imperative to provide shelters, food, and water for these lovely songsters. It is equally important to make sure our cats are kept indoors or that they wear a tiny bell on their collar when outdoors to help reduce the death toll of our feathered friends.

It has long been known that the benefits birds bring outweigh any damage they may do to crops, but as with any rule, there are always a few exceptions. Use of hoop-greenhouses over raised beds in the garden prevents disruption from problem birds while the crop is young and vulnerable. For strawberry crops, we use a similar idea but instead of plastic, chicken wire fencing is employed. The hoops are light and can be removed easily for weeding or harvesting.

However, we have had cherry trees in the past, and during ripening—birds may be surprised to find they are not as welcome as always. Our neighbor, Mary, was talking with us about the birds that continually robbed her tree. When we mentioned the use of netting for a barrier, she countered that in her experience, nets actually entangled the birds. Not wanting to harm wildlife, we searched all our resources for ways to reuse items in ways that will deter birds only during the harvest season. Here are some of our favorites:

Nearly everyone by now has received at least one of those unsolicited CD's from computer service companies. These are wrongly thought of as useless, however, strung up in the problem trees, they will rotate and reflect the sun. This shimmering effect tends to scare the birds off. One could apply this with old music CDs as well.

Defective videocassette tape may be reused in much the same way. Cut into lengths, they will shimmer and make noise, which is an affective deterrent to birds and other skittish creatures like deer. For more details, see *Around the Garden*.

Remember to take these deterrents down as soon as the harvesting is complete. The birds will return to their wonderful practice of eating harmful pests and their larvae.

Benefits
- Protect bird populations by keeping cats indoors.
- Keep pest and weed populations naturally controlled with birds.
- Protect fruiting crops from birds with defective videocassette tape or old CD's.

FURNITURE

When in the market for new furniture, check out used stores and the classified ads in the newspaper first. Often one can find like-new items for very reasonable prices. If you do purchase brand new items, look for ones with the greatest multipurpose potential. However, before buying new, take a good look at the furniture you have. Can it be repaired, reupholstered or refinished?

Wood furniture is easily refinished. Even small dents can sometimes be fixed. Starting with stripped wood, soak the dent with a wet sponge for several hours; then lay a metal bottle cap over the dent and heat with a hot clothes iron. Heat the cap with the iron but do not allow the iron to contact the wood. The wood will swell under the cap and fill the dent. Otherwise, use wood putty to fill larger dents and scratches. Finally, sand and stain, paint or varnish. It is as easy as that and all you need for tools are a few brushes, a couple of reused glass jars, and some sandpaper.

If the family has decided the old furniture just will not do anymore, there are still a few options left before taking it to the dump. Consider donating, trading, or selling the unwanted furniture.

Or take it apart and reuse pieces. For example, drawers with casters attached to their base are excellent for storage under the bed or to hold a heavy potted plant. Kitchen chair pads can be removed and used in a variety of ways. Use as a kneeling cushion for work in the garden or workshop. Keep one in the car for emergency repair work. Take and use as a comfortable seat when camping, to picnics, or at parks, anywhere there is bound to be hard benches.

Another option to purchasing new furniture is making your own unique items yourself. One need not be a woodworker with a well-equipped shop either. Wooden spools, commonly used for heavy gauge wire, can easily be made into tables. Shave off any rough areas. Sand, then paint or stain the spool—depending on the condition of the wood and your desires. Then apply felt or soft cloth on the base to prevent it from scratching floors. One could even attach small wheels for easier mobility. The top can be laminated, or covered with a piece of glass, or a mirror. Alternatively, simply cover the entire thing with a nice cloth. Either way, you now have a unique, solid table.

Simple desks, cupboards, and tables can be easily made from wooden doors. Hollow core doors can also be used, but try to use the finished edges as much as you can. Where you have open ends, you can finish off with moldings. A few of these would work well as an instant wall for concealing a furnace area or hot water tank.

Creating lamps is easy with the light fixture kits now available at hardware stores. Using almost anything as a base is possible. The only limit is your imagination.

Demolition projects around town can reap many rewards. One couple we know recovered a few pieces of marble and made a new floor for their outdoor patio. Pieces of concrete made them a lovely garden path. Another friend salvaged several old sewing machines and used the sturdy metal legs for matching glass tables. He also managed to secure all the major crossbeams for his cottage home for free from a motel unit under demolition.

We could not end this chapter without briefly discussing the wonderful world of pallets. Wood pallets are typically constructed of red or white pine, maple or oak. Despite the value of their wood, they are surprisingly easy to attain for free. Use them whole as instant trellises in the garden or take them apart and make them into boxes, benches, furniture, and much more. One enterprising individual constructed his entire barn from recycled pallets. Dave made a greenhouse several years ago that was constructed entirely out of crates recovered from a workplace's waste wood. Once complete, no one could tell by looking at it that the materials were recycled.

Benefits
- Save money by purchasing used furniture.
- Create storage space or plant stands out of sections of furniture.
- Create your own unique ambiance by making new furniture from recycled objects.
- Build furniture, barns, greenhouses or compost bins out of recycled lumber.

GLASS

41 billion glass containers were produced in the U.S. in 1992. In 1994, approximately 13.3 million tons of glass waste was generated in the US. Food and beverage containers made up 91% of this amount: the remainder came from products like cookware and glassware, home furnishings, and plate glass. A ton of glass produced from virgin materials creates 384 lbs. of mining waste. In Germany, 76% of soft drinks are in refillable containers; in Austria, 95% of mineral water and in Norway, 60% of wine and liquor are also refillable.

Because glass takes so long to decompose, the bottle thrown away yesterday could still be littering the landscape in the year 3000. Fortunately, today, all glass food and beverage containers can be recycled.

Most glass food jars produced today contain a minimum of 35% recycled glass. Industries are now creating innovative products from recovered glass such as blasting abrasives, construction aggregate, glasphalt (asphalt and glass mixture), reflective paint, pipe backfill, fiberglass, tiles, and insulation. The fiberglass industry alone, used more than 500,000 tons of cullet (crushed glass) in 1998; with as much as 40% recycled content in new product. Frost heave damage to highways and pavement is greatly reduced when cullet is used as an ingredient during construction, mainly because it encourages water filtration.

Of the 37 billion glass containers produced in the U.S. (64% clear, 23% brown, and 13% green.) in 1996, 90% ended up in the waste system. By the year 2000, over 26% had been recovered—yet due to industry consolidation and competition, the number of manufacturing facilities had declined from 127 to 59.

When preparing glass for the recycling depot, it is important to remember it should be rinsed clean. Remove the lid, if it is metal it can be recycled with cans. And remember; only bottle or jar glass is acceptable—ceramics and other types of glass will only contaminate the bin unless otherwise stipulated at your local recycling depot.

At the depot, glass is usually required to be sorted by color. This is because clear glass is the most valuable, whereas mixed glass is the least. Because the quality of mixed cullet is inferior, market applications are then limited.

Glass reuse has become a growing global activity. Refillable beverage containers are common in many European countries, with the Netherlands being the leader in glass recycling, reusing 80% of all its glass containers. In Denmark, 97% of all beverage containers are refillable. Many Canadian companies are now using the refillable option as well. Approximately 90% of The Beer Store's products in Ontario are in refillable containers. A winery near Toronto also employs refillable wine bottles. Dairies throughout the country are once again using refillable glass milk bottles.

Such global endeavors as these warm environmentalists' hearts!—especially when finding uses for glass around the home is not very easy. Articles have been written in a variety of magazines on exterior building walls being constructed of glass bottles mortared together. We heard of an old-wives-tale that said placing glass jars in the soil so that the open tops whistle in the wind will deter pests. We reluctantly tried it one year and found this method not only unsightly, but also completely unfounded. We do reuse jars for storing food and seeds, but because we can be clumsy, and glass is breakable, plastic is often preferable for storage containers.

It has been an exciting experience to witness glass recycling programs escalate very quickly in the last 10 years, and this trend will only continue. It is predictable then that other waste glass materials, such as light bulbs, may also be considered a valuable resource one day.

Benefits
- The energy saved from recycling just one glass bottle can light a 100-watt bulb for 4 hours.
- Making use of recovered glass saves 81% of the energy that would have been spent converting raw materials to new glass.
- Recycling glass reduces air pollution by up to 20%.

HANGERS

Clothes hangers can sometimes pile up on us, leaving our closets cluttered. Everyone seems to have enough and so giving them away to people is not always an option. There are a few places to consider. Dry cleaners will often be more than happy to take back their own hangers. Many thrift and second hand stores, churches, community halls, or shelters are in need of clothes hangers—so give them a call.

Hangers can also be reused around the home outside of our closets. We reuse our metal hangers in the garden as pins to hold row covers and plastic covers in place. Simply cut the two elbows out so that there are 6" long legs left to insert in the ground. Poke the legs through the cover or plastic, and push into the ground like a giant staple.

To make them into a dust wand, simply cut out the hook part of the hanger and straighten it out. Attach strips of recycled soft cloth on the end of the wire with a glue gun. The wand works very well for reaching ceiling and wall corners where those nasty cobwebs seem to accumulate.

While enjoying a campfire, why not reuse a hanger or two? Wire hangers can be straightened out into a long rod that can be reused repeatedly to roast wieners or marshmallows.

Benefits
- Returned hangers reduce costs for clothing establishments while avoiding waste.
- Save work in the garden by making your own pins.
- Create your own dusting wand or roasting sticks.

METALS

Scrap Steel

Do you have scrap steel lying around your yard, barn or field? This year might be a good time to take it in for recovery. Recovered scrap metal production results in a 97% reduction of mining wastes, 86% less air pollution, and 76% less water used. Annually, enough energy is saved by metal recycling in the U.S. to supply Los Angeles with almost a decade worth of electricity. Recovered steel is used to make a variety of products, creating employment along the way, and reducing the need to open mines in virgin lands. More than 90% of steel made in Canada will be recycled, and 50% of new products contain recycled steel.

Most loose scrap metal and large appliances may be recycled at your local landfill. Any item containing ozone depleting substances are often accepted at landfills where the refrigerant will be removed by the staff. Check with your landfill office on their particular policy.

Cans and Lids

Aluminum cans are recyclable. In fact, the recycling of aluminum saves 95% of the energy needed to make new aluminum. The energy saved from recycling a single aluminum can is enough to operate a television for 3 hours! Today, over 30% of the aluminum used in industrialized countries is produced from recycled metal.

When preparing cans for recycling, simply rinse them clean and remove any paper labels. After having her ankle sliced by a can lid that had gone undetected in the bag, Lillian began placing any sharp lids inside the can, then tightly pinching shut the top of the can to enclose it.

There are endless ways we have seen others reuse cans before recycling. Clean tin cans make excellent cutworm collars for transplants in all your gardens. Remove the label from the recycled can and cut out the bottom as you did the top. File sharp remnants so that fingers and transplants are not harmed. Place so that the transplant is centered inside the can and press the can firmly into the soil (about 1") to stabilize. Flower bulbs can be protected from moles in the following way: with a juice punch, press several holes in the bottom of the can. Bury the can in the ground with the bulbs inside, bottom down to provide a shield from those subterranean pests.

Use a large can as a double boiler in a pot of boiling water to melt wax. Or use in place of seedling plant pots, being sure to punch drainage holes in the bottom. You can hide the look of a recycled can by painting it or inserting in a wicker basket or clay pot. Alternatively, we have used larger cans as an outdoor ashtray for our guests by filling them with sand. We have also seen cans used as scoops and even hung in trees as a deer deterrent.

There are many craft projects children can make with recycled tin cans. (Be sure to file the inner lip so there are no sharp edges). They can be made into pen and pencil holders, a tiny drum-set or a steel drum—all which can be painted. Punch or drill holes along a penciled design, paint and then use the can as a candleholder; it will let out dots of light on the walls in the pattern you have chosen. To make a child's telephone toy, punch holes in the bottom of 2 cans and run a rope through them, tying a knot on the inside of the cans. When the string is stretched tight, words spoken into a can on one end will be slightly audible in the next.

You can use cans as a candle mold. Smooth sided ones will release the finished candle easier. Simply tie the wick to a stick lain across the top of the can. Fill with melted wax (broken crayons or old candles will do) and cool. Heat just the exterior of the can in a pot of hot water and the candle will slide out with some coaxing.

Here is a handy overhead light bulb changer for the vertically challenged or height-sensitive folk. Gather together the following materials: 10oz-soup can, a Styrofoam cup and a pole. Nail the can onto the pole and insert the cup into it. Gently place a bulb in the cup, just tight enough to grip the bulb. It is an excellent tool to use for changing hard to reach light bulbs without having to use a ladder.

We must not ignore all the money our families are missing if we do not take advantage of bottle and can deposits. Beer, soda, and juice cans are all return for deposit.

Aluminum foil

All aluminum foil products are recyclable, yet there are ways to make good use of them before taking to the depot. Reuse clean foil trays from takeout and prepared foods for freezing homemade pies, casseroles, and lasagnas. You can save quite a lot of money this way. These homemade meals make wonderful gifts for the elderly, bachelors, the busy, and the ailing. We like them for the days when we do not have time to cook a meal or when we are having unexpected dinner guests and would rather not pay the price of take-out.

Save yourself some money and time spent cleaning the stove by reusing foil pie trays to line your oven burner trays. Simply cut a piece out to accommodate the plug. These trays can be periodically wiped clean in order to extend their lives.

Aluminum foil sheets can be reused to cook with several times, prolonging their life. We have a system for this that makes the chore much easier. Place sheets in soapy dishwater, lay flat on the counter, wipe with a cloth, rinse well, and let set in the dish rack for a while to dry. This way the sheets remain flat and tear less easily. Reuse to line the bottom of the oven and ease the cleaning chore. Or line the stove burner trays instead. Hang strips and balls in your trees to scare birds from the fruiting crops. Alternatively, make into anti-cutworm collars for transplants in the garden. Before we started reusing foil, we went through several rolls a year. Now we rarely even go through one.

There are other sources of aluminum to consider recycling, such as lawn chair frames and window frames. Check with your local scrap metal dealer for details.

Benefits
- Recycling aluminum saves energy.
- Recovered metal is used to make a variety of products, creating employment and reducing the need to mine virgin ore.
- Protect transplants and bulbs from pests.
- Reuse cans for craft projects.
- Prevent physical strain by creating an overhead bulb changer from recycled materials.
- Save money by reusing foil products.

NYLONS

For women, many careers and social occasions often require that they wear nylon hosiery. Unfortunately, nylons are not made to endure in a world full of sharp edges that catch and create unwanted snags or runs. Some women stop runs by dabbing some fingernail polish on, which in effect glues the run together, preventing it from lengthening into a total blowout. Typically, this will only postpone the damage long enough to get through the day intact.

Hosiery is best hand washed in warm soapy water and hung to dry. Handled gently, they can last through several washings, but the fragile fibers of the material will eventually become brittle and break down. Sadly, they do not readily biodegrade and are not recyclable. Before you throw them out, consider all the things those old (but clean) nylons could still be useful for.

To use as a strainer in the workshop, simply stretch a piece over a container (i.e. a paint can), secure with a strong rubber band and pour through it to separate any lumps or impurities. A friend has used this same strainer idea for processing lump-free gravy, sauces, and jellies in the kitchen, and we have tried using them as a strainer for seed sprouting—which seemed to work fine.

In the garden, slip the foot part over ears of corn when the silks have begun to darken slightly, preventing insects from invading the ears. For best results, stretch it firmly over the ear and tie in a knot at the base. Alternatively, use an elastic band or twist tie to hold this ear-cover on.

When saving your own seeds, slip the foot portion over the seedpod after the seeds have begun to form. Tie the heel around the stem to safeguard against insects while preventing shattering and the subsequent loss of seed. Or, slip the foot over grape bunches to discourage bugs and birds.

Use lengths of the legs as very strong, yet gentle, ties to support heavy branches on over laden fruit trees. Incidentally, nylons tied to metal stakes, cages, wire trellis, or fencing can actually generate extra nitrogen into the soil, as long as the metal penetrates the soil. This is due to the static electricity created by the stockings, which is then conducted through the metal and into the ground, bringing nitrogen from the air along with it during the process.

The thick waistbands of stockings tend to be the strongest part of pantyhose and therefore are the best part to use in securing young shrubs and trees to a support stake. Waistbands work well as giant elastic bands for holding garbage bags on the can or bundles of newspaper together. Large square pieces of stocking laid over the vent on the back of your computer's hard drive will help to keep dust out.

The thick bands of stockings also work well in an emergency. Clark, an airplane mechanic during World War II, and a vehicle mechanic thereafter, tied his wife's stockings together one Sunday drive, using them as a temporary fan belt. That stocking fan belt survived just long enough to get the family safely to the next garage that was open—over 3 miles away! Of course, they drove *very* slowly!

Joanne used to make her own rugs by crocheting rags together. She found that nylons could be crocheted with a giant rug hook into one of the best bath mats she had ever had. The mildew resistant properties of the nylons keep the rug stain and odor free and with no absorbency, the rug dries out very fast.

If you are interested in potpourris, you may elect to try this idea. Stuff a section of pantyhose with aromatic dry herb mixes and tie or sew the end(s) closed. Attach to a coat hanger, or string, and hang in your attics, closets, and cupboards. A potpourri of sage, basil, rose petals, citrus rind and lavender is very pleasant. The herb mixes you choose can be for fragrance only, or for repelling insects, depending on the need. Cedar shavings (available at pet stores) repel moths and other flying insects. The bag can be beautifully decorated with a little ribbon or cloth.

On CBC Radio, a female caller relayed how she made nylons into what she called *ice-bags*. She cut the legs off pantyhose, filled them with salt and tied them to the roof of her house to aid in keeping it clear of ice. She claims that when ice does accumulate, she easily scrapes it off because it does not stick as it normally would without them. The only concern might be how the salt would affect any landscaping growing beneath the drip line of the roof.

We had a giggle over a letter we once read in a newspaper of another unique use for hosiery. An older lady cut off the crotch and legs of control-top panties to fashion a sort of halter-top to use as a bra. On a more serious note, another reader wrote how after breaking a rib one day, she was unable to wear a bra, so she tried the first lady's idea and was pleased that it worked. And yet another reader wrote in a similar idea that applied to mastectomy patients. Apparently, the breast form can sometimes slip out of a bra, which would not be a comfortable situation in a social interaction. The woman solved the problem by slipping a section of nylon over the form before inserting it into her bra giving it the grip it needed to stay put.

It just goes to show—never underestimate the extent of reuse ideas possible from any given item!

Benefits
- Extend the value of your dollar by straining impurities from liquids.
- Prevent pest damage and seed loss in the garden.
- Support fruiting branches and stressed stems.
- Make a temporary fan belt.
- Generate nitrogen in the soil.
- Create useful nylon bands.
- Save money by making useful crafts.

OVEN RACKS

Old oven racks can come in handy for many uses. Before considering reuse, run some sandpaper over them to remove any rust or metal barbs. For a nicer appearance, repaint with any color.

To make a handy shelf for your outdoor patio or greenhouse plants to rest on—use logs, rocks, or bricks to elevate the rack from the ground. It will allow excess water to drain where that water may be reused by another plant.

During harvest season, the same simple elevated rack can be reused to dry onions, potatoes, or garlic. Put by the outdoor faucet to wash fresh garden vegetables.

Use for drying your mud caked or wet hiking boots, by setting up a rack shelf and setting in the open air under protection from rain. While camping, these racks hung horizontally by cords on all corners from a tree, or some other sturdy structure, make a fantastic drying rack for clothes, towels, and even shoes. Things dry very quickly this way. Oven racks are handy when having an outdoor fire. Placed over the pit, it provides a handy cooking surface for pots.

Benefits
- Save money by reusing as drying racks and shelves.
- Prevent damage to pots and pans by employing as cooking grates for campfires.

PLASTIC RINGS

Do you remember hearing stories about how those plastic rings from a 6-pack of beverage cans could endanger wildlife? Picture this: an unsuspecting duck paddles through a littered waterway and finds herself entangled in plastic rings. Frantically struggling to get free, the bird only manages to tighten the binds of plastic and eventually dies from exhaustion, exposure, or starvation. Occasionally, the bird may free itself by chewing through the plastic, only to perish later from intestinal blockage. This sad event is not limited to heavily polluted waterways, but rather wherever human activity has occurred. After envisioning such a horrible thing, finding ways to keep these rings out of the landfill does not seem like such a chore.

Unfortunately, at the time of this writing, plastic rings are not recyclable at any depot we have visited. For years, we have ensured that any plastic rings that head for the trash can are thoroughly cut up so that wildlife cannot get entangled. After much research and questioning, we have found some interesting ways to reuse the rings.

Plastic rings can be used to support vines and fruiting plants. Staple or tie them together end to end until you have made a belt at least 4 feet long. Make wider belts by stapling sets side to side. Now you can hang on the rafters or shelves of your greenhouse to support your vines or fruiting plants in a hammock fashion. Attaching the ends to any vertical support allows mini-versions of the hammock to be used on individual fruits and vines around the garden as well. An ambitious gardener once used them to construct a fence for his garden. What he needed was not a strong barrier but one to deter the animals from his area. He attached them together by tying each ring with a piece of recycled string.

One ingenious young mother connected the rings with colored ribbon, tying bows at the connection of each ring. She used this as a stuffed animal hammock for her children's play area.

Once again, industry has risen to the challenge of the plastic problem—from bags to cups. Plastics, traditionally derived from petroleum products, have now been developed from corn in order to make them more biodegradable. The process converts the corn sugars into a substance called polylactide (PLA), which goes through fermentation and distilling processes before being formed into plastic beads used to produce most plastic products, including some fabrics. At the Winter Olympics in Salt Lake City, Coca-Cola used cups made from PLA plastics, which successfully composted in only 40 days. Naturally, there are details to iron out and it will take some time for these new technologies to become commonplace, yet the news is astounding. Just imagine; that instead of filling our community landfills with plastics, we would have large composts that nourished and beautified the city's parks and gardens. Truly a green environment!

RODS AND HANDLES

Old broom handles, tool handles, curtain rods and the like, can all be very handy to hold onto—no pun intended. The reuse of various metal and plastic handles is virtually endless as well.

If you are handy with tools, a long wooden broom handle can be used to fabricate a stick horse. Cut the handle to the desired length, and using sandpaper—remove any splinters or sharp edges. Trace two identical horse head shapes onto a sheet of thin plywood. With a jigsaw, cut out the shapes and attach to the handle with screws or nails. It can be easily animated with a bit of paint for the face and recycled yarn for a mane. These make great toys for visiting children and excellent gifts, as well. Lillian has fond memories of summers spent at her grandparents' farm playing on a stick horse made long before she was born.

Long handles can be used as hanging racks. Remove the tool head and cut off any remaining threads. They can be used in the workshop, garage, utility room, or garden shed to hang tools, rope, or cords on. With mounting hardware now available, it could be made into a coat hanger rod inside a closet space. It also makes for a good clothes-drying rack in the utility room for those items that cannot be put in the dryer. It stores away easily and when needed, can be supported by two chairs or hung from the ceiling with cord.

Hanging the rods in a greenhouse can be very functional. Placing a sheet of linen over a few rods hung across the ceiling will provide quite a bit of cooling shade for the wilting plants in the heat of summer. Rods work well in the garden as stakes to support tall and lanky plants. For bush beans make a support fence by pushing several of these rods into the soil in a row. Use wire, or string, to make a fence between the rods - wrapping around each rod as you go. Or, with 3 or more rods, construct a teepee with the beans planted in a circle around the base. When fully foliated, it will appear as a green cone of beans with no visible support.

To build a trellis for climbing plants, simply nail some horizontal rods across some longer vertical lengths in a ladder-like fashion. Make the trellis wider at the top by using increasingly longer horizontals. Keep a few inches of the vertical rods on the bottom free so you have some rod to pound into the soil. A similar ladder can be used as a chicken roost or ladder. The number, strength and length of rods used is simply determined by the needs of your project and the materials available.

Alternatively, cut the rods into any desired length and replace faulty chair and stool legs, or use as feet for homemade furniture or planters. Shorter lengths can be used as tie down pegs, or to replace handles on small hand tools. The limit here is your own imagination.

Benefits
- Prevent this resource from heading to the landfill and save money on material costs.
- Save time and money by tidying storage areas with hanging racks.
- Cool off greenhouses.
- Save money on garden stakes and support materials.

SOAP

We gardeners are stubborn; even in early spring you will see us pounding trenches in the barely thawed earth for our legumes! All joking aside, gardening is dirty work and the clean up of soil-stained hands can be a difficult chore.

Lillian's mother came up with a great way to reuse pieces of bar soap for this purpose. In a recycled container, save all the tiny pieces of bar soap that are too small for use. When you have several cups worth of the pieces, chop them up in a blender on low speed. Add a bit of water until it attains a pudding-like texture. Place in a recycled plastic container and use it to wash hands. Be sure to keep the lid on the container or the soap pudding will dry out. On the other hand, if it is a little wet, leave the lid off and air-dry for a few days.

We have successfully diluted 1 Tbsp. of the soap pudding in a bucket of very hot water for washing our truck. This same solution has been regularly employed to clean the garage and workshop floors at our home. It makes for an easy cleaning job, effectively removing most dirt and leaves a nice lingering scent behind. Similarly, this cleaning solution should work well on most other heavy cleaning activities.

Alternatively, place the soap chips in an old nylon sock. Press them down into the toe and tie it off. Hang this soap sock under the garden or house exterior taps where hand washing will occur. Simply scrub your hands with the sock.

Benefits
- Extend the value of bar soap.
- Decrease need to buy hand soaps, and building or car washing detergents by reusing.
- Reduced amount of packaging means less waste headed for the landfill.
- Fewer items to shop for results in fewer trips and less time spent shopping.

STYROFOAM

Before 1988, there was no recovery system in place for expanded polystyrene products or Styrofoam. Any step we can take toward recycling or reusing Styrofoam diverts this waste from our landfills where it can take more than 500 years to break down. Of course, the best way to prevent this waste is to avoid buying products that come in this type of packaging. Fortunately, fabricators are beginning to accept returned polystyrene for reuse, converting it into a variety of products. Among these are produce packaging, food service and industrial trays, office accessories, picture frames, piano keyboards, CD boxes, single-use cameras, and videocassettes.

Sadly, most food packaging is too contaminated for recycling. Perhaps in the future, foods could be contained in biodegradable materials that are readily broken down in composts. In the meantime, you can reuse styrofoam trays from non-meat food purchases in the kitchen. Filled with fresh baked goods or garden produce and covered with plastic wrap, these packages also make nice gifts. At our drumming workshops, where we serve refreshments, these are employed as small serving trays for snacks.

When it comes to reusing styrofoam packing chips, frustration can mount quickly. Due to static cling you soon find yourself speckled head to toe with the feather-light chips. Wiping the hands, chest and arms with a new dryer sheet can eliminate the problem. The sheet can be reused in the dryer, and in the meantime—no chips will stick to you while you work.

In our storage room, we keep a large box handy for packaging materials such as bubble plastic and Styrofoam chips. Clean food trays, egg cartons, and the foam that comes in boxes with electronics can also be contributed to this box. When packaging parcels for shipping, these items come in handy. Large blocks of Styrofoam are also useful in ponds and pools for easing stress caused by ice in winter months. We have quite a reputation for packaging materials with our friends, who pop by occasionally to use our supply for a package they intend to mail.

One year, while taking our annual potted plants out of their containers, we noticed there was an awful lot of unused soil at the bottom where the roots had not even reached. After some research, we learned that this problem could be remedied by filling the unused space with Styrofoam chips. For food crops, this may not be a good option, but the space can be filled just as effectively with sand or gravel. By reducing soil needs, we experience fewer shopping trips, save money and have less packaging to throw out. The plants also seem to have less mould and fungal problems, most probably due to increased drainage.

Many drum-kit players place a blanket or pillow inside the bass drum to control the sound. A correspondent with *Modern Drummer Magazine* wrote about using Styrofoam chips instead. He filled the drum for very quiet gigs, but reduced the chips as required for different volume levels.

Statistics show that the average North American family consumes 40 dozen eggs each year, the equivalent of 1kg of polystyrene packaging foam! We are avid gardeners and for a time helped run the family market garden and seed supply farm. There, we were always looking for something in which we could plop one more seed. Egg cartons work wonderful for starting seeds. First, remove the lid and any attached flaps. Then poke a hole in the bottom of each cup for drainage. Fill with soil, moisten thoroughly and plant the seeds. Flip the lid over and reuse as a tray to catch any drainage water. Cover the whole works with a recycled bread bag, cut open so that it lies fully flat, until the seeds germinate. When transplanting, use a spoon to scoop the seedlings out of the cups. If a paper carton has been used, simply cut the cups apart and plant so that the edges of the cup are below soil level. The paper will soon decompose, allowing the roots to grow through.

Egg farmers are often looking for cartons. Some grocers selling farm fresh eggs will accept empty cartons and return them to the farmer when he comes back with the next delivery. Or you could try phoning advertisements of farm eggs for sale in the paper, to see if they need any.

In the craft room, reused cartons are useful to organize small items. Or make a unique gift box by taping two lids together along one edge. Tape both sides of the edge, so that they hinge open like a lidded box. Wrap the box with recycled fabric or wrapping paper and tie shut with a nice, recycled ribbon. Because gift boxes can be reused many times, they are a gift in themselves, especially for those who find the chore of wrapping presents difficult. For more wrapping ideas, see *Holidays*.

Day-care, kindergarten, and pre-school classes are often looking for donated egg cartons. Used in making all sorts of crafts, the children have a wonderful time while the school saves money on supplies. Aspiring artists can use the cartons as disposable multi-chambered palettes for mixing paints.

Alternatively, store golf balls in them! One suggestion we found on the Internet was to cut off the tops of the cartons to use as trays for cookies and such for children. An elderly neighbor, swore that they work very well for making ice cubes—though we have never tried this ourselves. In the kitchen, inverted egg cartons make perfect holders to support your taco shells while they are stuffed with filling. Canadians can mail their polystyrene cartons to the Canadian Polystyrene Recycling Association (7595 Tranmere Drive, Mississauga, ON. L5S 1L4). The company will also recycle the boxes incurred from all the mailed packages.

Benefits
- Any step taken towards diverting this resource from our landfills is positive.
- Save money by making unique gifts.
- Use as packing material for shipping.
- Reduce wasted potting soil.
- Control drum-kit volume.
- Help egg farms keep costs down.
- Reduce need for seed starting trays.
- Use as craft material.

THE GARDEN

As avid gardeners, we find many reuse ideas are easily implemented, and a lot of dollars can be saved while growing our own nutritious food.

Barbecue

Use metal lids and bases as exterior plant pots. Barbecue wheels attached to large plant pots allow for easier mobility. The same idea applies to any other sturdy wheel from toys or machinery. A lawnmower with the engine removed makes a great low-to-the-ground dolly for moving heavy potted plants around.

Garbage cans

Often, garbage cans have cracked or broken edges. Simply cut the can down and try some of these ideas: Use as a recycling bin, storage for potting soil or soil amendments, storage of garden tools, drill a few drain holes and use as a plant pot, or use as a liner to build a small pond in the garden.

Teapots

Old teapots make wonderful planters for a unique decor. Put fine sand in a 1" layer on the bottom of the teapot. Then place your plant in and fill with good soil.

Alternatively, use as a potsherd in gardens for frog or toad homes. Lay on its side; bury only by an inch, leaving the opening above ground.

Plant Tags and Labels

Recycle the metal end from frozen juice cans. Punch a hole in each lid. Write the name on the flat surface in large, neat printing with a permanent marker. You can either tie on to the plant with recycled string, or nail to a stake to identify the plant, or crop. These can be reused for many years.

Cut clean plastic containers into 2-inch wide strips. Taper one end so that it is sharp and will be easier to insert in the soil. Using a permanent felt marker, write the plant name on the side.

Venetian blinds make the best reusable plant labels. Cut them to size being sure to taper one end. Dave finds that scissors will cut PVC; tin snips work best for aluminum blinds. Mark the plant name with a permanent marker.

Garden Hoses

If there are only a few small holes and cracks, an old hose can be made into a drip hose for landscaping. Simply drill a line of tiny holes along the length of the hose before using and cap the end. Protect outdoor electric cords by slitting the hose lengthwise and putting the cord in.

Replace bucket handles by slitting the hose lengthwise and inserting the wire handle; tape closed.

Hoses can be used to support trees. First, pound in a sturdy stake near the tree. With a length of wire long enough to wrap around both the tree and the stake, slip a piece of hose onto the wire where it will contact the tree. When the wire is tightened to support the tree, the hose will keep the wire from cutting into the bark.

Lay short pieces of hose along the ground in earwig infested areas as traps. Early in the morning before the day gets warm gather and immerse in a bucket of soapy water until all the pests drown. The hose pieces can be reused many times for this purpose.

Video tape

Scare birds away from trouble areas with defective videocassette tape. Cut the tape into 2' lengths. Hang on tree branches or push sticks into the ground every few feet around the area, and tie a few tape lengths to it. The slightest breeze causes the strips to shimmer. This is only temporarily effective, as the birds will eventually lose their fear of it. Put out when protection is needed (i.e. cherry ripening season) then remove, so that the birds can carry on their vital contributions in the war against pests.

Waxed and Styrofoam cups

Use as starting pots for seeds. Simply cut a few drainage holes in the bottom. Fill the first inch with recycled Styrofoam chips, sand or gravel, and top with soil. Then water the soil well, letting set before sowing seeds. For similar propagation containers, see *Styrofoam*

Hair

Put bags of hair in tree—softball size, to deter deer. Contribute hair to the compost.

Garden Pots

Gardening pots tend to accumulate, and for the beginning gardener, these pots can be a treasure. Yet, in order to prevent disease or damage, they must be thoroughly cleaned after each use, then dried, stacked, and stored somewhere. Once a good supply of cleaned pots, ready to use for seed starting or cutting propagation has been secured, most gardeners are at a loss as to what to do with any surplus. Of course, asking gardening friends to see if they need some is always a good idea. Sometimes you can find a local nursery owner willing to reuse the pots to cut costs.

The larger pots are reusable for plants, like tomatoes, before being put in the garden. They can be used on outdoor decks for flowers, herbs, and other container gardening. These pots come in quite handy for holding a transplanted shrub or tree until a site is prepared, or until it has reached the desired size. Buried in the ground, the pot will contain the roots so it is easily dug up and moved later. Mint, being so invasive, can be contained this way by preventing the runners from spreading throughout your garden. Those extra pots are much appreciated when relocating to a new home, and you wish to bring along some treasured garden plants.

Clay pots are attractive, but tend to dry out plants quickly due to the clay's lack of moisture retention. Get around this by placing a potted plant inside that is in a slightly smaller plastic pot, using the clay pot as a dressy facade only. Broken clay pots can be used to make frog homes in a damp area of the property. The ideal site will supply leafy plant growth for shelter. Press the edges into the soil and pile a bit around the sides to hold in place.

Even harder to deal with are the thin, plastic, multi-cell packs. They arrive at a nursery in sheets and are typically separated into sets of 2, 4 or 6 cells for the home gardener's transplants. These are difficult to wash and being so flimsy they break or bend easy, making storage or reuse difficult. That results in very few gardeners keeping this type of pot. Try to avoid purchasing transplants in these cell packs.

However, a new program by the Canadian Polystyrene Recycling association (CPRA) is aiding gardeners in keeping cell packs and molded plastic trays out of the waste system. They are now sent to the CPRA plant in Mississauga, Ontario where this former waste is converted into a resource. The plastic is made into pellets and sold to plastic manufactures, where products like cell packs, CD cases, molded plastic trays and office accessories are produced. Many garden centers are beginning to participate in the collection process, but so far, the program is only active within Ontario due to shipping costs of these bulky plastic items. We can only hope for the success of this program and its expansion across the continent.

Soil Blockers

Despite increasing reuse and recycling efforts, there is still a lot of unnecessary plant pot consumption and waste happening in North America. We believe Ladbrooke's *Soil Blockers* may just be the answer to this expensive problem. Initially we saw these on a TV show, *Gardening Naturally*. Made of metal, these tools make grow blocks out of soil, eliminating the need for plant pots all together. This saves money not just in the cost of pots and the time put into cleaning and storing them, but also in soil.

The first stage is the 3/4" propagation cube maker. It presses 20 cubes at once with a slight concave depression for the seed to germinate in. If a seed does not sprout at this stage, you have only a minimal amount of soil to discard. Next up is the 2" blocker that punches out four blocks with a square hole for a 3/4" prop cube to fit into. They also come with short pins to create a seed sized hole for sowing directly into the 2" block or longer pins for the propagation of cuttings. Next up is the 4" blocker that creates a square hole for the 2" block to fit into. Each of these hand-operated blockers is set up so the size previous fits in its depression. Because the soil blocks are small and square, they take up much less room than pots do. The roots, being oxygen-trimmed, are ready to spread out when transplanted, rather than wrapped around the bottom of a pot. This makes it possible to visually check the roots of a plant to determine when blocking up to the next size is necessary.

Imagine the time and labor saved using these block makers. Now think of all the pots you will not have to buy, store, and clean annually. It is our hope that nurseries will start using these as well. There are stand-up versions available for the serious gardener or commercial operation that can squeeze out dozens of blocks at a time.

Benefits
- Create recycling or storage bins out of broken garbage cans.
- Make frog homes out of broken pots.
- Make plant tags and labels out of blinds and plastics.
- Turn old hoses into drip hoses, earwig traps and tree supports.
- Extend the life of buckets and the comfort of their handles by reusing pieces of hose.
- Keep birds away from fruit trees with defective videocassette tape.

- Turn hair into a soil amendment or deer deterrent.
- Purchase fewer pots and less potting soil.
- Save money, prevent packaging waste, and reduce your own waste.

TIRES

Approximately 3.8 million tons of tires were generated in the US in 1995. Only 17.5% of these were recycled—not including those that were either retreaded or combusted at a waste-to-energy facility.

E-magazine had an article about a tire-burning incinerator that produces electrical energy, which although intriguing, raises obvious concerns regarding the emissions. Thankfully, tires do not necessarily have to be burned to keep them out of our landfills. Landfills now accept tires for recycling, but first the rims must be removed. Check with your local landfill for their policies on tires.

Innovative environmentalists have proven tires can be regularly reused as a building material for houses, fences, gardens, roads, and dams. By far one of the most impressive accounts of reusing tires is for a roof covering sealed by tar. Requiring very little maintenance, they are proposed to last up to 2000 years! Tires can be cut and placed, rounded side out, to line horse stalls and are easily employed in landscaping projects like foundations, ponds, and waterfalls.

In beautiful Grand Forks, we have seen several places along the two rivers where old tires been have linked together and filled with gravel or rocks in an effort to prevent erosion of the bank. Planting indigenous wild flowers, grasses, and water plants inside the tires will aid in keeping a more natural appearance, promote wildlife, and aid in stabilizing the tires. Sand dams, designed to prevent soil depletion due to wind erosion, have been successfully used for years. The materials for tire sand dams are as much as 90% less expensive than concrete dams, but are also more labor intensive and time consuming.

A tire can be reused for entertainment as well. Young and old alike enjoy tire swings, hung in the shade of a big tree or from the rafters in a barn. Attached horizontally to a building, they make good hoops to shoot balls through. In fact, one Saskatchewan farmer placed a tire up high in a tree for his son to practice throwing a football. Dave grew up with a tire-swing in the backyard. When his dad was teaching him to switch-hit for baseball, the tire swing doubled as the practice target. This helped him develop a strong left-handed swing. Many parks have employed tires in various ways to construct activity areas for youth.

Plant trees inside large tires and cover with mulch. Loop a piece of soaker hose around the base of the tree for easy irrigation. The tire will help to protect the young tree from weather extremes. The mulch and soaker hose will maintain the proper moisture needs for the tree. Tires are particularly useful as miniature raised garden beds for squash and other heat-loving plants such as melons. The black rubber acts to absorb the sun's heat, warming the roots of the plant.

Meat, bones and grease can be composted safely by reusing tires. Build a stack of several tires and fill with soil. Bury the meat waste in the center ensuring it is covered deeply. Then place a piece of recycled chicken wire or wire screen over the soil and cover with a thick layer of mulch. The mulch will maintain the moisture level and eliminate fly problems, and the wire will keep rodents from burrowing in. Because the tires are black, the pile will stay hot enough to kill unwanted microorganisms. Leaning tires can be corrected by pounding in a sturdy stake for support. Leave the pile to break down for 2 years, and then introduce it (in layers) to the garden compost where it will compost further.

Our favorite invention for reusing old tires is the manufacturing of irrigation soaker hoses. They are made entirely out of recycled tires and are flexible so that they are easy to place on garden beds. They are readily cut to fit the size of the beds and are very durable— we have been using the same soaker hoses for over 8 years. Soaker hoses are well known for their water reduction capabilities—using less than 1/3 the water of an overhead sprinkler. Also, it seems practical to quench your food crops and not the pathways and borders where weeds will thrive if watered. The less weeding to waste time on—the better. Incidentally, World Water Day falls around mid-March, so this may be as good a time as any to consider this water saving technique. It is a great way to show support for a great product made from recycled materials.

Tires need not make the home look trashy. We have seen people paint their projects with beautiful results. Some choose solid colors. Others have solid backgrounds with nature or floral scenery on the foreground. We drove by one home that had their children take part in painting with stick people, big yellow suns, and the like all over their compost bin made of tires. It looked beautiful in their garden behind the children's play area. With some creativity, and a little fun, the look of your home can actually be uniquely improved by reusing what was formally destined as trash.

Benefits
- Prevent erosion along waterways.
- Construct dams, homes and roofing using tires.
- Create activity areas, swings, and sport training tools.
- Save money by building your own miniature gardens and compost piles.

TRASHY BITS

This chapter is a collection of assorted reuse ideas for some common items. As there are numerous things you can do with any one item, this is only scratching the surface of the concept of reusing. The only limits are your imagination and creativity.

Baling Twine

Most especially on farms, baling twine can be found strewn in fields, along fencing, in the garden, tucked into nooks and crannies, wrapped up on sticks, or just balled up together. Taking the time to cut each loop into a length and tying each length end to end is a great sitting-on-the-porch-in-the-shade job. By this, we mean it is a good excuse to go sit and relax while busying your hands. Once you have a good length of rope formed, wrap it around a stake or stick. In no time you will have a good collection started. It can be used to construct pea and bean trellis-type fencing and to use in place of rope anywhere, except where a lot of strength or beauty is desired. Once the climbing beans cover up the string, visibility is no longer an issue.

Batteries

With recent technology, industry has thankfully improved the efficiency of the recycling of batteries. Nickel cadmium and other rechargeable batteries are recyclable. Call 1(800) 8-Battery for a drop off location near you that takes rechargeables. These batteries have better performance than in the past, and are now commonly found in cell phones, laptops, cameras, and hand tools.

Many retail and electronics stores that sell batteries will take back your old ones. Lead acid batteries from automobiles are accepted at landfill sites for recycling as well. Alkaline batteries can now go in the trash, as they no longer contain mercury.

Belts and Watch Bands

Belts work well as strong straps to support young trees being that they are strong enough to support, and yet soft enough to not cut into the bark. Cut into desired lengths to glue under items like desks, dressers, tables, or anything that may scratch the floor surface. Cut to the desired size, a belt can make a good pet collar. Leather watch bands can be reused as cupboard door mufflers. Simply cut into small pieces and glue to the inside corners of the cupboard. This makes the doors close more quietly. With watches being very low priced, it can be tempting to simply throw them in the trash and consume another one. Consider replacing your battery or buying a new band, before opting for the wastebasket.

Caps & Corks

Painted bottle caps can replace missing board game pieces—or use as a small mixing bowl in the shop for glue or paint. Corks can be used as a key chain float or as fishhook covers in the tackle box. Construct a corkboard by gluing corks onto a section of plywood. Corks will compost over time.

Computer and Audio disks

Reuse junk mail CDs as reflectors for vehicles traveling at night. Simply nail them, shiny side out, on fence posts and mailboxes or wire them to gates. Our good friend, Brian, reuses CDs as coasters. Backpackers can use them as emergency signaling mirrors. See *For The Birds* to use discs to keep birds from fruit.

Construction Materials

Landfills are increasingly accepting asphalt, asphalt shingles, bricks, concrete, and re-bar for recycling. Some landfills charge a fee to cover equipment costs. Policies may vary from one landfill to another, but usually pieces must be no larger than one square meter. For information on recycling wood, see *Trees Please*.

Coolers

Keep groceries cool during transport in summer months by keeping a cooler in the vehicle. They can be donated to youth groups that do outdoor activities. The lids make an excellent outdoor picnic tray. Many cooler lids have convenient drink holders on the under side. Use to store toys, beach or pool ware, or as an activity box with art and craft supplies. In the yard, they can be used as a mini pond, a birdbath, or water basin for thirsty pets. For container gardening on the patio or for storing garden harvests in the cold room, a cooler is a good strong device too valuable to discard.

Cups

Because paper cups are not recyclable, every time we use one, we contribute to the growing landfill problems. When coated in wax, these little conveniences do not decompose easily. The process of manufacturing Styrofoam or polystyrene is particularly damaging to the environment. It is a difficult substance to recycle, and even the depots that do accept it—will not take food containers.

People can choose to bring a reusable coffee cup to work, rather than consuming disposable cups. If this is not an option, try reusing the same cup for the day by rinsing out after each use. This is especially practical when its only been used for water. The benefit of reusing cups is keeping office supplies and waste disposal costs down.

However, both paper and foam cups can be put to use as transplant pots for your gardening. See *The Garden*.

Ceramics are not yet recyclable, but abused cups still have some use in them. Old cracked or stained mugs make beautiful pots for small houseplants and transplants. Many an office has one on the desk as a pen or pencil holder.

Crayons and Candle Wax

Create your own candles. Melt bits of wax in a coffee can over a pot of hot water and run through a fine sieve to remove any debris. Reheat and melt again while you prepare desired length of wicking string. Use crayon pieces to color to your liking. Dip string into melted wax many times until the desired thickness is met. Or with reused cups as molds, pour multi colored layer candles and simply peel away the cup when cooled. Rub your kitchen and workshop drawer edges with wax so that they open more smoothly. For ideas on using wax to construct fire starter logs with newspaper and egg cartons, see the *Newspaper* and *Paperboard* chapters.

Curtains

Before discarding, consider donating window curtains to thrift shops, second hand stores or shelters, or sell them yourself at a garage sale. If they are too tattered and worn to be of value, there is still reuse potential in them.

Utilize as a drop cloth around the shop, under a child's kitchen chair, or wherever crafts and painting take place. Reuse light-colored, sheer nylon, or polyester window curtains as row covers in your garden. Line the trunk of your car to protect the carpet from a dirty cargo.

Use old, clean shower curtains like a smock to protect clothes when painting or other messy art projects. Cut into large rectangles and secure them to the shoulder seams of the kids' shirts with clothespins. A shower curtain under the tent will protect the floor in wet weather. Extend the life of the shower curtain by using a hole-punch to create a new hole for the rings to go through.

Fish Netting

Old fish netting is often available from coastal fishermen. Use as deer fencing around gardens and crops. You can easily stitch any holes with recycled rope pieces and sew layers together until you reach the desired height. Netting has also been employed to protect bird-ridden crops such as berries or fruit trees.

Ironing Boards
Old ironing boards make excellent potting benches and are often found at garage sales. They are just narrow enough to fit nicely in most greenhouses. They will fold up and store away when not in use, and the holes in the ironing board provide ample drainage for transplants. When we have a lot of plants hardening off, this bench can be placed anywhere convenient.

Light Bulbs
Light bulbs work well when repairing tears in clothing—especially socks. The bulb's shape is perfect for forming the tear to the body's form, and because it is glass—the needle is easier to control. Those who are crafty, tell us light bulbs are as commonly used as molds in paper-mache.

Paints, Pesticides, Solvents, and Fuels
Many landfills accept these at on-site reuse programs or for recycling. Call your landfill for information on their policies.

Pens and Pencils
Choose mechanical pencils over wooden ones, when looking to buy, as they are refillable and reusable. Pencil shavings stored in a film canister makes an excellent emergency fire starter for backpackers. Refillable pens of many shapes and sizes are also available in most stationary stores.

Pharmaceuticals
Many pharmacies will take back any unused prescriptions for safe disposal. Never dump them in the toilet or down the sink. Pharmaceuticals in our water system are wreaking havoc not only on our own health, but in wildlife systems as well.

Plastic Eggs, etc

Children's toys and gifts often come packaged in an assortment of plastic containers—most commonly egg-shaped. Filled with aquarium gravel, rice or lead shot, these make a good sounding musical shaker. Dave has made several of these over the years and uses them both live and in the recording studio. The larger eggs from pantyhose work well to package small gifts in. These look really nice when permanently decorated. Lillian once had a very large egg with an Easter bunny design glued onto its outer surface. It was so nice, we used it in an Easter supper table display. The following year it was filled with chocolates and given to a child as a gift.

Propane Tanks

A friend of Dave's has constructed some very creative ornamental pigs out of old propane tanks. Unfortunately, not everyone has the welding equipment or expertise to do this kind of project, but propane tanks are now recyclable at many landfills.

Pots and pans

Pots and pans, dishes, and most other kitchenware can be donated to charities and thrift stores or sold at garage sales. They can be reused as toys in children's sandboxes or for playing at the beach. Camping results in far too many bumps and bruises for good kitchen equipment to be taken out. Many campers reuse older pots because it does not matter if the bottom is burned from the open flames of a campfire. Consider replacing any broken handles with those from a second-hand pot to extend the life of an otherwise perfectly good utensil.

Shakers

Reuse plastic herb, spice, or Parmesan cheese shakers in the kitchen. They can be utilized in the garden by filling with soil amendments like kelp meal or rock phosphate. Use to sow crops like grass, clover, carrots, and parsnips, which are commonly planted fairly thickly.

Swing Set Greenhouse

Often found for free or very cheap at garage sales, old swing sets can have their life extended—as a green house. Remove the swings and cover the frame with 6-mil clear plastic. Use cut coat hanger pieces to pin down outer edges, then cover the entire edge with dirt for added security. For extra insulation, pile leaves or hay on the edges in the fall. One creative fellow attached a zippered door from the recycled family tent for the entrance, using duct tape to attach it.

Umbrella

Hang the stripped umbrella skeleton upside down from the showerhead and use as a foldable drying rack for clothes.

Telephone wires

Multi-colored strands work well for arts and crafts or easy plant identification in the greenhouse. Use to tie tomatoes or other climbing plants onto their support systems.

Thermometers

Unknowingly, North Americans contribute tons of mercury into the landfills by way of old thermometers—the largest source of mercury in our waste systems. Consumers can combat this problem by choosing to purchase non-mercury thermometers. Old or broken ones should be taken to your local public health department where proper disposal can be ensured. It is equally important to take old compact fluorescent bulbs to a hazardous waste facility, as they also contain mercury.

Three ring binders

Cut out the ring section. Open the rings and cut the upper half of the rings off—using bolt cutters or a hack saw, for instance. Drill holes in the flat surface so that you can hang, nail, or screw the panel to a wall. Hung on the wall, they can be used to hang keys, hats or coats.

Toys

One technique parents can try is to exchange toys with friends and family rather than buying more new ones. This will ease the cost for the parents, while providing variety for the child. Another approach is putting half of the toys in storage then switching them when the children begin to show signs of boredom with their present playthings.

Garage Sales are popular ways of buying and selling used toys. Donating toys to a daycare, hospital, or doctor's office is a very good option as well. Whether you are donating or selling the toys, try to repair the broken ones as well. A number of years ago, we met about an older couple who bought used toys, bringing them home to their workshop to repair and repaint before giving them to poor and needy children. A modern day Santa's workshop!

When you are purchasing new toys as gifts, parents recommend keeping the quality of the toy in mind. The longer the toy lasts, the longer it will be enjoyed.

Video and tape labels

Save the unused labels that come with video and cassette tapes. Use them as shipping address labels and on recycled envelopes. To recycle envelopes, just attach the label over the original address. Use to label storage jars and containers of homemade preserves, craft supplies, and children's toys. Identify packages in your freezer, particularly if you are reusing plastic yogurt containers.

Wine box canteen

As an alternative to the age-old canteen, consumers can now choose from a variety of bag-type water reservoirs. Light weight and collapsible when empty, there are times when just one will not do and several can be costly. If you drink alcohol, consider buying a *box* of wine next time you are purchasing. Not only are you reducing the packaging that those 5 or 6 bottles would have created, you now have a bladder from inside the box that can be reused as a backpacking canteen. Wine bladders are made of lightweight plastic that is very tough. They hold about a gallon of water with rubber stopper spouts, and when empty, they fold neatly away into the size of a wallet. Simply remove the drained bladder from the wine box, soak in a weak solution of bleach water for a few hours, then rinse and use.

Insulate by making a sleeve out of an old foam pad. Cut into a rectangle twice as long as the pad, fold in half and duct tape the seams, leaving the top open for the bag. Often on day hikes, we will fill the bladder with ice pieces before topping off with water, or freeze the bag overnight, half full of water to keep its shape. Then fill with cold water just before leaving. This way nice cold water is available for most of the day—a pleasant indulgence on a hot hike.

Recently, wine boxes and bottles have become returnable for deposit. Be sure to take advantage of this.

PART TWO

AFTER SIGHT

Sadly, many visually challenged individuals are required to renew their prescriptions regularly. Those old glasses do not have to be discarded. There are many eye wear establishments that collect used glasses and distribute them to those who are not so fortunate. These programs can have enormous impact on both the reduction of waste in the landfill system, and in aiding the visually impaired all over the world.

For Example, in 1988, Lens Crafters and Lions Clubs International teamed up to create a program called *Give the Gift of Sight*. This program set a record by receiving 784,000 donated pairs of glasses in 1992 alone. Volunteer optometrists and technicians travel to developing countries to provide free eye examinations and distribute the eyeglasses where they are needed the most. This astounding example of people's generosity is really only a small percentage of the needs for an estimated 1.5 billion people worldwide.

Many of us can take a look in the attic and in our junk drawers and find a few pairs of unused glasses. We, too, had several pairs of old prescription glasses, and honestly have no idea why we have kept them. No longer do we have to imagine our hard-earned money going into the trash can; now we can donate them to someone who could really use them. Perhaps the individual will be able to learn to read, attain employment, or clearly see a loved one's face for the first time.

Benefits
- Reduce waste in your local landfill system.
- Aid visually impaired people all over the world.

CLEAN WALKING

North America is so very beautiful. Our family has learned to appreciate its nature more and more each time we venture out for a hike. Sadly enough, though, not all of us appreciate our pristine land, letting garbage and unburied waste lay where it falls. We have come upon many piles of garbage in the wilderness and along trails where people have obviously dumped their waste on purpose.

Trash in the wrong place can be devastating and dangerous to humans and wildlife alike. As a youth, Dave once released a duck from entanglement with a fishing lure. It had the hook through its beak and some line wrapped around its head. In Lillian's youth, a young boy she was swimming with received serious injuries when he stepped on a broken beer bottle. And let us not forget that wildlife can be habitualized into eating from our waste. Most bears that become a nuisance have all too often gotten that way due to the careless actions of a human being. It could be a hunter leaving the remains of a kill, a rural gardener with open, fragrant compost, or campers leaving food out. All these events can turn any animal into a doomed one.

After complaining over such disrespect, we soon realized that by walking by the mess—we were behaving no better. When we first began the practice of collecting waste, it was a little embarrassing as we imagined that people were thinking we were so down on our luck that we had to collect bottles from the ditch. Then Lillian was stopped twice and thanked for cleaning up the area, and we realized that our imagination had gotten away with us.

Whenever we go out for walks, we bring along a few plastic bags and a pair of gloves. Picking up garbage is an ugly job, so we only pick up the non-organic waste—any food or pet waste can be scraped aside using a stick and covered with a log or a pile of leaves. With one person using a stick to pick up the waste and another to hold the bag open with gloved hands, no one needs to actually touch any garbage. Often, much of what we gather is recyclable, and we usually come home with some returnable bottles and cans for the depot.

In the summer, when the temperature is warm enough for swimming, we clean up under water. Whenever we go to a lake or river, we bring along the snorkeling gear to remove waste and any dangerous objects. Submerged logs are prime targets for fishing lures to get snagged upon. Late summer brings the lowest water levels and this is the perfect season for cleaning up this deadly mess. We found this chore can have other rewards. We once found a very expensive diving mask in 20 feet of water.

This project can be turned into a more fun and rewarding event if the whole family or a group work at it together. Your group can take great pride in having played a part in making this world just a little bit better! It was very encouraging to see another couple doing the same thing while we were parked at a highway rest stop. They were almost to the point of rappelling off the edge of a rock bluff to retrieve some trash.

It only takes a few minutes of our time, yet takes nothing away from the enjoyment of our outdoor activities. In fact, it gives us more incentive to get outside more often. Our reward comes from returning to cleaner trails and waterways that are filled with healthy wildlife. There is a certain amount of pride felt when you have helped make an area so pristine. The next trail-user will experience a better, safer world because of this simple effort. Numerous studies prove that tourists return to an area primarily for its cleanliness and greenery. In this era, where the economy has come to rely more on tourism, cleaning up is truly a benefit for the community.

Benefits
• Improve your community's economy by keeping it pristine for tourists.
• Decrease pollution by collecting valuable recyclable resources.
• Keep medical costs to your community in check by removing waste from waterways, parks, and trails.
• Reduce incidences of animals foraging or perishing from trash.

DIET

Although the topic of diet choice is not exactly a reuse or recycling subject, our diets do affect resources and contributions to the waste system. For instance, in the year 2000, the average individual Canadian ate 10 oz of meat daily. To put this in perspective, producing 10 oz of beef requires 35 times more water than 10oz of potatoes. The livestock feed industry requires 64% of our agricultural land—feeding animals, which release 60 million metric tones of methane gas into our environment. And 50% of our water consumption is due to some phase of meat production.

The vegetarian diet has strong historical roots, which not only stem from geographical implications, but also for the love of life— or rather, a reluctance to destroy it. Many religions, especially those that believe in the transmigration of souls, think it is immoral to eat meat (i.e. Hinduism, Buddhism, Jainism, Doukhobor, Seventh Day Adventists, and Trappist Monks). Costing only a fraction of meat-based diets, vegetarianism also eliminates concern over hormones, antibiotics, toxins, bacteria and other undesirable elements often found in meat.

Clearly, this diet choice shows both a compassion for animals and the preservation of our natural resources. Even reducing the number of meat-based meals in your diet relieves some of the stress on our medical system, and improves your health and budget while bettering the environment. And it places you in the company of many famous vegetarians. K.D. Lang, Albert Einstein, Paul McCartney, George Bernard Shaw, Julia Roberts, and Mahatma Gandhi are all vegetarians of some degree.

Our journey into a more vegetarian based diet went from being quite strict vegetarians to where we are now, which is about an 80/20 vegetable to meat ratio. We have never felt more vitality in our lives before, and our cost of food has been dramatically decreased.

Benefits
- Reduce impact on natural resources.
- Increase health and vitality: relieves stress on medical system.
- Decrease consumption of hormones, antibiotics, and toxins in the diet.
- Save money by purchasing less meat.

ENERGY USE

We decided to include this subject in Trash Talk for two reasons:
1. To waste energy is no different than throwing your hard-earned money out with the trash.
2. The production of mass utility energies creates pollution and the destruction of wilderness areas. Pollution is simply put—trash in the wrong place.

Heating and Cooling

The average household spends 1,300-2,400 dollars annually on energy—44% of that being for heating and cooling. The fossil fuel consumption for one home during a winter season is responsible for more emissions than 2 cars in one entire year! To make the home more energy efficient you do not have to be a contractor or spend a fortune. Simply keeping the heating and cooling systems clean and regularly replacing the filters will improve their efficiency, while ensuring that the indoor air is cleaner. Be sure to check the ductwork for any leaks that are easily repaired with (you guessed it) duct tape.

Keep the heating system's thermostat set to around 68° or 70° for the most efficient and comfortable indoor air temperature. Reduce it to about 65° at night, before retiring to bed. When you are planning to leave the home for the day or a number of days, turn it down to 55°. For every degree reduced, up to 5% of winter heating costs is saved. When having several guests over for a gathering, consider turning down the heat a few degrees. The increased body heat will make up the difference.

Used to maintain heat and protect from cooling, proper insulation can save 30% on the home's energy usage. Holes and cracks around windows, doors, light fixtures, outlets, and walls should be sealed. For exterior doors, replace the weather stripping and doorsills as soon as they show any signs of wear.

If your budget allows, consider replacing single pane windows with the more energy efficient, double paned type. Low-emission coatings reduce energy exchange through windows by up to 75%, and they can reduce the need to cool or heat the home considerably.

Installing plastic covers on the exterior windows during the winter months is an option. If removed carefully, the plastic can be reused for numerous years before the UV rays make it too brittle. A simple wooden frame can be made to attach the plastic with butterfly nuts holding the frame in place. Requiring only a minute or so to install or take down, this eases the annual chore of stapling to the exterior window frames then ripping it off again in springtime. It is true that utilizing plastic window covers will hinder the view, but does not affect houseplants or compromise indoor lighting in the least. If it bothers you, an alternative would be to put the plastic up only in the very coldest of months. It requires a brave soul to take on such a chore in the cold of winter, so we usually have ours done before too much snow has hit the ground. Generally, we do not plastic the windows that provide our most cherished views.

Keep the south facing windows very clean during the cold seasons to make optimum use of the sun's warmth. Using heavy curtains, especially during the evening, will reduce cold drafts but be prepared for possible condensation on the windows in the morning. Open them during the day and have a fan going to prevent any mold or mildew problems caused by moisture build up on the glass. In the summer, keeping drapes or shades closed until the sun no longer shines directly on them will greatly reduce the cooling needs of the home.

Fans reduce moisture problems and circulate warm air in the cold season, maintaining an evenly heated home. By creating a breeze, fans also cool the air in the summer. Properly employed, they can eliminate the need for an air conditioner entirely. Refrain from setting the air conditioner colder than usual—it will not cool the home any faster. Keep TVs, lamps, and other heat producers away from air conditioners to prevent overworking the machine.

House exteriors that are dark in color absorb 70-90% of the sun's energy. White and shiny surfaces reflect the energy of the sun away, contributing to urban glare, which contributes to the warming of the Earth's stratosphere and are no longer recommended as an exterior finish. So depending on your energy needs, a new coat of paint can make quite a difference in your energy bill. Ideally, if one's house could change colors like a chameleon, it could adapt to each season's requirements.

Clean Energy

Converting your existing home or building by incorporating clean energy can bring many benefits. For instance, converting to air-source heat pumps can cut costs of heating by 40%. Some government programs exist that aid in the financial cost of conversion, especially for low-income families. Energy improvements may also qualify the home for *Energy Efficient Mortgages*, which take into consideration the higher income to debt ratio over the long term. It is true that the cost of new technology is expensive, however, costs involved in wind-generated electricity have reduced by 5 fold since the 1980's. Other alternative energies are expected to become more affordable as demand and the level of technology escalates.

With this in mind, those constructing a new home or refurbishing an old one can consider solar, wind, or geothermal energies.

Landscaping

Landscaping improves the energy efficiency and monetary value of your home, and the beauty of your community. Landscaping is not necessarily limited to the ground level either. Gardens can be found on balconies and on rooftops (known as green roofs), which can reduce the amount of storm runoff going into our sewer systems and help clean the air. Landscaping and green roofs reduce urban glare, and city residents and workers experience a more positive disposition and better health due to these little park-like areas. Trees and shrubbery provide shade in the summer and wind protection in winter, alleviating energy costs.

Lighting

For those who are constructing or refurbishing a home, there are several ways to make use of natural lighting. Choosing light colors, installing skylights, and the use of mirrors are some options to consider. Use appropriate window coverings to make the best use of lighting in each situation.

Rather than illuminating the entire room, try to focus lighting where you need it—with under the counter flourescents or track lights. Dimmer switches and 3-way bulbs are favored for the control they provide over the intensity of light, thereby reducing glare and creating mood settings. Using one 100-watt bulb, rather than two 60-watt bulbs, provides the same amount of light (lumens), but uses 15% less energy. Purchasing long-life bulbs, which can last 5 times longer than normal bulbs, further reduces contribution to the landfill.

Better yet, invest in compact fluorescent lighting (CFL) bulbs for every socket in your home. CFLs have a life of about 10,000 hours, which is 10 to 20 times longer than traditional bulbs. Though CFL is not as harsh a light, a 25W CFL bulb will give the same amount of lumens as a 100W incandescent bulb. And because they need to be changed less often, CFLs are especially useful in hard to reach areas. When used for continuous exit sign lighting, CFL bulbs will last up to 2 years, saving money while reducing maintenance costs. Choosing these bulbs decreases time spent shopping, and reduces the amount of packaging and spent bulbs sent to the landfill—and less shopping trips! Also worth mentioning is that conventional light bulbs put out a lot of energy as heat, not light. Some bulbs (i.e. halogen) can actually create enough heat to start a fire. Christmas tree lights can now be replaced with cooler bulbs, decreasing the risk of a dried out tree catching fire. Outdoor Christmas lights, typically 7 1/2 - 9 watts, can be replaced with 5 watt bulbs—cutting power use by about 40%. Using a timer to control Xmas lights to operate between 7-11 PM is also a good idea.

Eight decorative outdoor gas lamps in one year can use the equivalent energy as that of an average home over an entire winter season. Refitting these with outdoor CFL bulbs is possible, but a cold-weather ballast may be needed to prevent dimming problems due to cold temperatures. Photo cell units and timers should be used to turn off the lights automatically. Make a dent in your energy use by employing timers (while away from home or for coming home at night) and motion detection lights—useful in areas where forgetfulness is a regular occurrence, or in entrances.

It is important to mention that stand-by power is a wasteful energy abuser. We are talking about stereos, VCRs, and microwaves that operate clocks or displays even when turned off. Unless the time display is necessary for that room, those lights are ever so slowly sucking money out of your wallets. We plug these electronics into a power bar with a shut off switch.

Keep an eye on your local retail store shelves for the latest energy-saving device or innovation. Industry is constantly improving to meet the new consumer demand.

Appliances

According to the *Canadian EnerGuide Appliance Directory*, the worst appliance consumes 1,011 kilowatt-hours of energy per year, while the most efficient available (in 2003) consumes only 184 kilo-watt-hours. That is a difference of 827 kilowatt-hours, or about $66.17 annually. With this in mind, it is easy to visualize the savings over the life of the appliance. Maintaining appliances regularly and keeping them clean will increase their life incredibly. Whether or not to repair your old appliance before recycling, is a decision that should be based on how efficient your model is. Typically with purchasing new appliances, the investment is paid back in a very short time. Choosing to run major appliances after 11 PM or before 7 AM (instead of during peak periods of energy use), reduces the drain on the power companies, preventing power outages. This method works particularly well for people on shift work schedules.

When purchasing new, choose those that have the longest warranty, are the most durable, and have the best energy and water efficiency for your dollar. Canada's EnerGuide label (provided by the manufacturer) is found on most appliances to show the comparison of energy use between models. Energy Star labels identify the most efficient models.

The laundry room is responsible for up to 26% of the hot water energy use in North American houses. To reduce this percentage, choose proper washer settings and use warm water only when necessary. Many detergents work effectively in cold water. We only use the warm water setting for the very dirtiest of clothes and have never used the hot water setting. It is important to note for those in the market for new machines, that front loading washers consume only half the water as conventional models. They also force more water out during the spin cycle, resulting in 50-65% less energy required for drying.

Our dryer is usually set on medium or low and is placed on a timer so unnecessary drying will not occur. The low setting also prevents heat from damaging clothes. In dust and bug-free areas, outdoor clotheslines are another drying option. Be sure the line is not located near road traffic in order to prevent pollution landing on your clean clothes. One can also install a redirecting dryer vent that allows the warm air back into the house, rather than piping it outside. The problem lies in homes that cannot handle the extra moisture in the air. Run consecutive loads in the dryer to take advantage of the accumulated heat. Keeping the lint catcher clean allows for proper air circulation.

Dishwashers can consume almost 14% of the home's hot water costs. Run the dishwasher only when full. Running the hot water tap for a few moments before starting saves the dishwasher from having to heat the water further before washing. Instead of using the heated dry cycle, air-dry or towel-dry dishes, using even less energy while preventing the possible warping of any plastic ware. Towel drying also frees up the washer for the next load, in case you have accumulated a great pile.

Gas cook-stoves with electric starts are said to be the most efficient. Watch for yellow flames, which indicate inefficient burning. Cooking areas that are clean reflect the heat better. Covered pots and salted water boil more quickly. Matching the size of pan to the size of the heating element also makes greater use of the energy.

Approximately 9% of the home's energy use is needed to operate the refrigerator. New energy-efficient refrigerators can save $35-$70 annually. There are also models with auto-moisture control features that reduce the need for the heater to kick in and battle condensation. Condensation problems are increased by improperly covered foods. Placing warm food in the fridge will increase its energy use, so be sure to allow leftovers to cool first. Thawing food in the fridge is not only safer, but helps reduce refrigeration costs. Be sure to organize the fridge and freezer contents well. Money flies out of those doors while you search for that last cold beer.

The most efficient settings for combination units are 40° F for the fridge compartment and 5° F for the freezer. Solitary freezers should be set at 0°F. It is important to defrost freezers regularly—a quarter inch of frost is enough to decrease its effectiveness. Place refrigerators and freezers away from direct sun, dryers, furnaces, and heating vents. Vacuuming the coils found on the backside at least once a year, and checking the seals regularly will keep the appliance running efficiently.

The Dutch—leaps ahead in waste management, are charging a tax on new appliances, which is primarily used to pay for research on finding ways of reusing every little bit of the machine. Many appliances, even the old ones, contain volumes of recyclable metals. To find a place that accepts old appliances near you, check in the Yellow Pages under Appliances. Old refrigerators that have been gutted-out and stripped of their workings can be turned on their backs and buried, becoming a mini-root cellar. Be sure the door remains a few inches above ground. On cold (but not freezing) nights the lid should be opened 1/4" for ventilation. One farmer put in a screened drain hole and slightly tilted the fridge so it would drain properly. Inside the fridge, the produce is usually stored in straw, sawdust, or sand.

There are many little things one can do in the kitchen to use energy more efficiently. For instance, we dry-roast almonds in a frying pan and immediately after—melt butter (for the same recipe) in the same pan, reusing the heat already accumulated. rather than using more electricity or another pan. Although this requires a bit of planning ahead, it ultimately results in less dirty dishes. When making cookies, we choose to use two baking trays. When one goes in, the other is being prepared and is ready to replace the finished tray without wasting energy in between batches. When flavor merging is not an issue, baking more than one item at a time can conserve energy considerably (baking potatoes with breads or casseroles). Incidentally, reusing heat and moisture that is created from cooking helps bread dough rise into lovely, light loaves.

Hot Water

Hot water accounts for a significant amount of energy use—14% of the average home's energy consumption, in fact. For those who are looking at purchasing a new tank, look for models with side or bottom cold water outlets—these are the most efficient. Consider putting in a solar water heater if southern exposure is available. Modern solar tanks can last over 20 years and prevent about 50 tons of carbon dioxide from entering the atmosphere. They reduce the consumption for heating water by up to 80%. Positioning the new tank in a warm area (furnace or laundry room) and away from potential drafts, will keep it from working harder than it should have to work.

Insulating your hot water tank can pay for itself in energy savings in as little as 4 months. A tank wrap kit can be purchased for around $20 at hardware stores. Go one step further and insulate the first meter of the hot water pipe to further reduce heat loss. Simply using insulation can save 15% in water heating costs—more if you have a poorly positioned, older tank.

There are ways to reduce the energy required to heat water in our homes. Begin by setting the water heater to 125°F—anything higher incurs energy loss. Drain the tank twice annually, sediment buildup decreases efficiency. Remembering to turn the heater down, or off entirely when leaving on vacation, can also save considerably. *On Demand* hot water systems consist of a mini water heater located under the sink to reduce the loss incurred in long distance piping. In a conventional system when a small amount of hot water is needed, gallons of heated water remains unused in the pipes afterwards. Alternatively, heat exchangers can be a good choice. These systems utilize hot water going down drains to preheat incoming water destined for the hot water tank, by way of copper tubing. This system requires no moving parts, power draw, or storage tank; yet, it results in a 30-60% decrease in water heating costs. Investment in either of these systems would pay itself off eventually.

Running the tap, as opposed to filling the sink, uses a lot more hot water. During the winter, leave the water in after a bath and it will release both heat and moisture into your home instead of in the drainpipes. Why should we let the hot water warm the sewer rather than our homes? During the summer, it is better to drain it and open a window to get rid of the warm, moist air.

On a smaller scale, using kettles to heat water consumes less energy than either a microwave or a stove. Homes heated with wood stoves can use the stove top to heat water, make tea or soup, and cook breakfast cereals.

Cleaning

You may have noticed that we recommend organic cleaning products throughout this book. By replacing the toxic products around your home and office with natural ones, you can actually save energy. Many products contain petroleum: a product with a sad history in energy requirements, and environmental and health risks throughout its production and transportation processes. Thankfully, many companies, such as Melaleuca, have created effective all-natural cleaning products.

Benefits
- Extend the life of appliances by using proper settings and maintenance procedures.
- Support industry improvements and green energy with your shopping dollar.
- Less energy consumed results in reduced pollution.
- Save money in energy costs.
- Reduce Urban Glare and storm runoff.
- Decrease contribution to landfill systems.

EVERY LITTLE BIT

The concept of zero waste is to see every little bit of our waste as a resource. We encourage you to take this concept to an extreme in your home. For instance, even floor sweepings, counter crumbs, vacuum contents, pet hair, and dryer lint are excellent contributions to the compost.

Saving vegetable and meat debris and cooking water for soup stock (as discussed in *Organic Waste)* is only going so far. Rinse out all the opened food containers, (i.e. soup tins) and add to your stock. When condiments like ketchup bottles and salsa jars empty, rinse them out and add them as well. You can even rinse out cooking pots and contribute that to the stock. One of our dog's favorite treats is peanut butter gravy (created when the jar is rinsed) poured over his supper. Mostly, you can use cold water for rinsing out jars and cans, but we use hot water to get greasier foods like peanut butter or tahini out of jars.

There are often a lot of soap and body products in use around the home and you can save quite a bit by using every little bit of them. When a bottle seems empty, there are several applications of the product left. Inverting bottles will allow the last of the product to flow to the top where it is accessible. If you have not tried this, now is a good time to start. Bottles with pumps are sometimes the worst culprits for wasted product because the tube does not reach to the bottom. It is amazing how much can be retrieved by doing these small, cost-saving steps.

You can go another step farther and rinse soap and hair product bottles as well. Fill the jugs about 1/8 full of water, replace the cap and swish around a bit. That liquid is then ready to employ elsewhere. For instance, liquid soap jugs can be rinsed, with the water being used for washing chores such as floors, walls or vehicles. Shampoo and hair conditioner bottles also provide several more treatments, when rinsed in this manner.

There is still enough toothpaste in the empty tube to do several brushings. Use a pair of sharp scissors to trim the bottom and one side off of the tube. This allows access to the inside. Simply use the toothbrush and wipe as much as you need to brush your teeth. Other tube products, such as antibacterial cream, can be emptied using this same method. We have had as many as one dozen applications retrieved from an *empty* tube of antibacterial gel.

Using a razor-knife, carefully cut open hand and body lotion bottles. Make a wide access by removing most of one side completely, leaving a box like structure behind. There are often at least 6 hand lotion applications in bottles that otherwise appear empty!

When you look at the amount of product that used to be put in the trash, this small chore will become easy to incorporate in your daily life. Over time, you will save quite a bit of cash because there are fewer shopping trips, you have made full use of the product, and you have less trash to throw out. Once people start, it is easy to become inspired to extend this idea to many areas in your daily life.

Benefits
- Make full use of your dollar by extending the value of products in your home.
- Save money on purchases that are less frequently needed.
- Reduced packaging means less waste headed for the landfill.
- Fewer items to shop for results in fewer trips and less time spent shopping.

HOLIDAYS

North American trash increases by 25% during the holidays, which equates to 25 million extra tons of garbage going to the landfill. Christmas is not the only gift-giving celebration, but it is responsible for more waste and consumption than any other holiday. The principals of many of the ideas that follow can be applied to the holiday of your choice.

Wrapping

Christmas, birthdays, and anniversaries are just a few of the occasions we celebrate with all sorts of surprises in beautifully wrapped packages. When we recognize the incredible waste that has accumulated at the end of the holidays, the urgency to do something about it becomes apparent. While most wrapping paper is recyclable, few consumers realize that the shiny paper that contains foil is not recyclable. We recommend that consumers avoid this type of wrapping all together. However, we found a great way to reuse it by making Christmas tree ornaments. Tiny blocks of wood or Styrofoam, or hand made cardboard boxes can be wrapped in the shiny paper. Decorate with a little ribbon and glue on a hanger (available at most craft stores). Alternatively, glue a loop of ribbon to hang the ornament.

Saving wrapping paper and ribbons from gifts you receive and reusing them for next year's gifts is easier said than done. In the excitement of opening a present, it is not on everyone's mind to do it gingerly. Try getting creative with recycled wrapping paper, maps, sheet music, posters, children's drawings, art, comic strips, and bits of ribbon and bows. Fabric scraps also make wonderful wrapping for gifts and cover the lids of preserves quite decoratively with a little ribbon. If every Canadian wrapped only 3 gifts in re-used paper or reusable gift boxes and bags, it would save enough paper to cover 45,000 hockey rinks. To learn about creating a wrapping kit, see *Paper Tubes*.

Reusable baskets, plastic containers and jars, can all be used for packaging up a gift. But why not make the wrapping a part of the gift? New dishcloths, scarves and towels are just a few ideas to use as unique and useful wrapping.

To avoid the unwrapping stage altogether, there are now different sizes of decorative gift bags available that are sturdy enough to reuse many times. Although there is an initial investment, these bags do reduce both waste and consumerism while encouraging reuse. We once received such a beautifully designed bag that we could not bear to store it away. Instead, it was hung for years in the back room to collect paper recycling.

To go a step further, we have also saved and permanently wrapped boxes with recycled wrapping paper. Just like the typical Hollywood gift when all they do is push aside a ribbon and lift a lid— same idea. If you do not have a box with a lid, make one. Tape the flaps of the box closed and with a razor knife cut off the top 3" or so. Now you have a lid but it does not fit over the box because they are both the same size. At the top corners of the box cut V slits so that when the edges are pulled together and taped, they make the top of the box smaller, allowing the lid to fit over the box. Now wrap the box and lid separately being sure to overlap the edges, securing with glue. Then finish by tying a ribbon or thin colored rope in a cross pattern around the box.

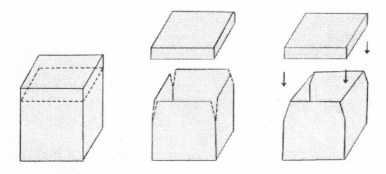

These re-usable boxes are especially appreciated by the elderly and disabled who have difficulty wrapping or unwrapping gifts. By doing this simple step, you will most likely influence others to do the same.

Cards

Greeting cards are recyclable, but they can also be reused in several ways before taking to the depot. Years ago, our friend Roxanne came up with an excellent cost saving idea for cards. By carefully saving all the cards her family received throughout the year, she had a collection of cards available for any occasion. Because most greetings are written on the inside of the back page, the front is often left blank. Cut along the fold to separate the two. Depending on the image, you can do a vertical or horizontal fold creating a new card or, unfolded, use like a postcard.

Making cards into gift tags is a very fast and easy craft. Cut sections from old cards and punch a hole where it will not disturb the picture. Then write on the blank side and tie on to the present with recycled ribbon.

Save as many gift cards as you can to make unique gift boxes. Take 4 cards of similar theme and trim to make them all the same size. Place a fine line of glue along the edges of the inside of each card to seal them closed. Laying the cards face down in a row, being careful that none are upside down, begin securing the boxes sides together with clear packing tape. Close the sides into a box shape and tape the last seam from the inside. Trace the box shape onto a piece of cardboard to make the bottom. After cutting, test it for fit in the box and trim if necessary. Cover with recycled wrapping paper. With one hand inside the box, glue the edges of the bottom one at a time, holding in place until set. Tape along the bottom seams to increase the strength. Because a lid is difficult to make, we stuff the top with colored tissue paper to conceal the contents. It is quite easy to attach handles to this gift box as well. Punch 4 holes, 2 on each side, near the top of the box. Tie two pieces of matching cord or yarn through the holes for loops and the box becomes much more appealing and more easily transportable. The boxes, one of a kind with a different scene on every panel, make excellent gifts.

A collage of cards can be laminated with plastic or contact paper to fashion place mats. Decorate the edge by gluing on some ribbon or cloth. Children can really have fun with this project and look forward to seeing their place mat come out during the holiday season.

By reusing these items before sending to the recycling depot, the home-crafter saves quite a bit of money on gift boxes and tags. The boxes go on to be gifts themselves as they can be reused for many years. Saving shopping time is only one of the additional benefits of this simple craft. Reducing consumerism also reduces the packaging waste destined for the landfill.

Gift Giving Ideas

Before you go shopping for a gift, consider making one. As active gardeners and cooks, our cupboards contain quite a variety of dehydrated and preserved foods. People really appreciate a jar of homemade jam, dried tomatoes, or chutney. They are easy to decorate and add a nice personal touch. Homemade baked goods—especially those that freeze well, make great gifts as well.

Alternatively, instead of a physical gift consider giving a service or experience. For instance, could the recipient appreciate a housecleaning, babysitting or lawn service? Tickets to a play, theater, or performance can bring a lot of pleasure to someone who may have little time to give to themselves. Gift certificates to a favorite restaurant, bookings to a drum workshop or yoga class, whatever you think they might enjoy. Giving these kinds of gifts improves local economies, while the recipient enjoys a unique experience.

As choosing something for a friend or loved is not always easy, gadget and gimmick gifts are often bought as a last resort. Singing ties and coffee cups of various anatomical shapes are not useful everyday items and often, after unwrapping, are put into storage—forever. Why not give an influence to better ecological practices? Nearly anyone could use an environmental gift of some sort in or around the home or office. Following is a list of ideas to start you off:

- A low flow shower head attachment.
- For the backpacker or camper: newspaper logs.
- For the apartment dweller: a book on balcony gardening or a worm bin.
- An automatic thermostat control that turns the heat down at night.
- Membership with an environmental group or magazine.
- For the Gardener: a subscription to an organic gardening magazine or a membership to a seed organization. Bird, bat, and butterfly houses, baths, and feeders make excellent gifts, as do interesting seed varieties.
- For the wood worker: patterns for bird, bat, and butterfly houses, worm or compost bins, or home improvement ideas.
- For the older home: a draft *cozy* (placed in front of doors to prevent drafts) and tubes of caulking or weather stripping.
- For the new parent: diaper services, cloth diapers, babysitting, housecleaning, or landscaping services.
- Brazil or cashew nuts: because they promote a living tropical

rainforest.
- Trash Talk book(s).
- Choose items that are powered by solar or rechargeable batteries.
- For the homeowner: a tree, compost bins, or worm bin.
- For the coffee lover: a reusable filter and whole organic coffee beans.
- Dimmer switches or compact fluorescent bulbs.
- Hot water tank and hot water pipe insulation kits.
- For the vehicle owner: tune-up or tire rotation services.

Christmas Trees

Every year sparks a debate among consumers regarding the purchasing of a real or artificial Christmas tree. Real trees are biodegradable, and when chipped, make good mulch for landscaping. Tree farms support life, provide jobs, and improve the environment as well. Unfortunately, many tree farms use pesticides and fertilizers. Artificial trees, although initially costly, save money over the long term. On the other hand, they are composed of petroleum products and cannot be composted. Consider donating unwanted artificial trees to someone you know or a charitable organization—instead of sending them to your landfill. Homeowners can try growing their own trees for harvest by planting quick growing white spruce or blue spruce trees themselves. All that is required is time, the occasional pruning, and watering.

Instead of investing in a harvested tree that is already dead, many people buy a live one. Evergreens fill the home with calming, therapeutic fragrances, and you can reuse them year after year for a one-time investment of anywhere between $30-$170. Shopping for that perfect tree annually is no longer necessary and you will save money every year thereafter, depending on how many years you reuse the live tree.

When choosing a live tree, avoid any that show signs of stress, such as brown spots or wilting. Live trees are heavy—so start with a small one, which tends to experience less shock when transplanted. It is important to pick a species of tree that grows naturally in your area, and to consider picking a different one with each purchase—to increase diversity and support wildlife. After the holiday, you can plant it in the yard or donate it to a business, school, or park.

Live trees that are brought indoors do best when kept in a well-lit spot, but away from direct sunlight or heat sources, as they prefer a cooler temperature. Water lightly every day; once the holiday is over set the tree outside in a sunny, sheltered spot to protect it. Keeping the tree indoors for too long will result in new growth that will freeze when moved back outside.

Once outside again, pile snow, sawdust, straw, soil, or sand around its base to help insulate it from temperature extremes and provide some moisture retention. If this is not possible, the tree will need a light watering every week or so. Also, try to keep the tender branches fairly clear of heavy snow build up, or you may experience damage due to breaking.

To keep reusing the same tree, supply it with a wooden container that allows for some expansion and contraction. Drill and screen drainage holes in the bottom of the container. Place the box on a tray with wheels for easy transportation. In season, bring the tree into a well lit, cool room (like a garage) for a few days before bringing inside the home. In the spring, light judicial pruning of the new growth (the pale green tips) is all that is necessary.

When you are ready to plant the tree permanently, it is important to pick a site where the roots will not disrupt pathways, and the foliage won't become inconvenient, usually about 20 feet from fences, gardens, or buildings. Be sure of the size and shape the tree will become before deciding on a location. In the autumn, spade the entire area (4 to 5 times the expected size of the root), and mix in one-third compost to the native soil. Transplant in the spring being sure to remove the pot or burlap from the root ball before burial. Lay down landscape fabric and mulch heavily with wood chips to deter weeds. Be sure not to over-water your tree—once a week in dry spells should do it. In addition, these outdoor trees can still have lights strung on them for the holidays.

Benefits
- Save money and reduce the consumption of wrapping, cards, gift tags, and Christmas trees without compromise.
- Encourage reuse and environmental ethics during the holidays by influencing through example, rather than preaching.
- Ease holiday wrapping struggles for those who find it difficult.
- Family craft projects with reuse techniques provide a chance to spend time with each other.
- Reducing shopping results in less transportation related pollution.
- Combat air pollution with a live Christmas tree.

INDOOR AIR

The human body can deal with most infections and natural organisms, but chemicals and fungi can cause havoc within our systems. Scientific studies have revealed that nearly 33% of new office buildings have air quality problems. The most common pollutants found in buildings are benzene, formaldehyde, radon and trichloroethylene—known to cause skin and eye irritation, headaches, appetite loss, eye and throat problems, drowsiness, nervous and psychological problems, respiratory problems, and even cancer. Location has a lot to do with indoor air pollution levels. For instance, buildings near high traffic roadways are likely to have more of those kinds of pollutants present in their indoor air.

Indoor air pollution does not discriminate; it can thrive in any of today's standing structures, including our homes. Plastics, paints, glues, and even carpets contribute to what has been called the *Twentieth Century Plague* or *Sick Building Syndrome*. Electronics and new furniture emit pollutants and carpets dissipate fumes into the air up to 12 months after installation. Particleboard, chipboard, and other glued boards are particularly guilty of releasing toxic gases. Concerned consumers should research before purchasing furniture or home products regarding the levels of Volatile Organic Chemicals (VOC) they emit in the air.

Polluted air can also result from fireplaces, wood stoves, dust and dirt, toiletries, pesticides, cleaners, and polishes. Test furnaces regularly for emissions, replace filters regularly, and install carbon monoxide detectors. Consider replacing the chemical cleaners used around your home with more Eco-friendly ones. As there are more allergy sufferers than ever before, it is now considered politically correct to refrain from wearing personal fragrance products. In fact, an increasing number of companies and doctors' offices are banning fragrances out of respect for those with allergies.

Due to poor ventilation and air circulation, mould and mildew can hide inside bathroom or kitchen exhaust fans, furnaces, and along windowsills. A common allergen that can affect people who thought they did not even have allergies, moulds, mildews and fungi is highly toxic, causing respiratory infections and chronic fatigue. Sterilize the areas using a cleaning agent able to combat moulds and bacteria. Employ tiny brushes, such as reused toothbrushes, to get into the tight spaces.

Since NASA officially recognized that living plants reduce sick days due to pollution-fighting qualities, many businesses are bringing greenery into their establishments. Plants, well known for their ability to cool the air, will help reduce air-conditioning costs in the summer. They release oxygen and moisture, alleviating the typical, dry, stale nature of indoor air. The presence of plants around the home and office provides a comfortable ambiance creating a more attractive place for your guests and employees.

Many varieties of plants are available, providing you with a choice of shapes and sizes. Variegated leaves, stems with striking colors, and various blossoms and foliages will make collecting fun. Several university studies have found that evergreen plants emit a therapeutic fragrance, so, why not throw in a little aromatherapy as well? You may experience a more pleasant and healthier environment to live and work in.

The requirements of a houseplant can vary as much as the microclimates can vary within each room of a building. For instance, one room may have several areas with available space for plants, but each area will be unique in the varying sun and draft exposure, temperature, moisture levels, types of pollutants, and human or pet traffic patterns. Therefore finding the plants suitable for your room is rarely a problem with all the different varieties available. Occasional watering, pruning of dead or unwanted growth, and potting-up when it outgrows the container are the only chores involved in keeping plants. By taking a few cuttings and placing them in a glass of water, you can propagate new plants easily and inexpensively. Change the water every few days to keep it fresh. When roots form around the stem, transplant into a pot of soil and water well.

The research we have done revealed this list of commonly recommended varieties that need very little care and can flourish under fluorescent office lighting. In our home, each room hosts as many as eight of these plants listed here. Do not let the Latin family names scare you—your local nursery will be able to help you find what you need:

- Azalea
- English Ivy
- Pot Mum
- Bibiscus
- Gerbera Daisy
- Philodendrons
- Bromeliade
- Golden Pathos
- Pomaettia
- Coleus
- Jade Tree
- Sansevieria
- Diefenbachia
- Janet Craig
- Silver Queen
- Dimcame
- Madagascar Dragon tree
- Spider Plant
- Dracaenna
- Orchid
- Weeping Fig
- Tradescantia
- Peace Lily
- orchid

Benefits
- Reduce health risks.
- Employers will notice a reduction in sick days.
- Reduce air-conditioning costs and combat dry and stale air.
- Greenery in the office creates a pleasing environment for workers and patrons.

INFANT AND FEMININE HYGIENE

Waste consultants, *Franklin Associates*, determined that 6.5 billion tampons, 13.5 billion menstrual pads, and their packaging were sent to American landfills in 1998. Consider the packaging alone; you have the box itself sealed in a layer of plastic and then each individually wrapped tampon has its own disposable applicator. The average woman uses as many as 15,000 disposable menstrual products in her lifetime (about 600lbs of waste).

Sadly, the waste from sanitary products is not limited to landfills. *Beach whistles* are not toys—they are the plastic tampon applicators that commonly litter ocean beaches. The Center for Marine Conservation found over 170,000 beach whistles along American coasts in one year. These objects are not only disturbing to vacationers and tourists, wildlife will often confuse trash with food and pay for this mistake with their lives.

Menstrual products are not the only sanitary products that are contributing to the waste problem on our planet. On average, a 2-year-old child has bestowed 5,289 disposable diapers into the system. Diapers contain many of the same toxins found in pads, and yet for long periods of time every day our delicate baby's skin is exposed to them. Rashes are often due to synthetic fibers and toxins irritating their tender bottoms. Often oblivious to the true cause, parents rush out to buy soothing creams or powders for the infant. It is a grand cycle, if you are in the business!

Many so-called sanitary products are comprised of synthetic and inorganic cotton fibers. More than 50% of the cotton grown in North America is inorganic or genetically modified. Cotton uses 25% of the world's insecticides, 10% of its pesticides, and 2.2% of its cultivated land. Some products also contain tree pulp that is treated with various acids. Both cotton and tree ingredients are processed using a variety of herbicides, pesticides, defoliants, and bleaching agents. Menstrual products also contain harmful residues, carcinogenic dioxins, surfactants, perfumes, waxes, and plastics. In 1980, a study done by the US Food and Drug administration discovered that tampon fibers imbedded in the vagina's membrane result in higher risk of infection, cancer, toxic shock syndrome, hormone disruption, birth defects, and other health problems.

There are reusable alternatives available that are more economical, safer, and much more comfortable. Among these are cloth diapers, menstrual cups and cloth pads with organic cotton alternatives for the discerning consumer. Some shy away from reusable sanitary products because they worry about the cleanliness of reusing them. We believe it is simply a matter of looking at the issue from another perspective. Consider underpants; we do not throw them out with every use – do we?

One day, Lillian was discussing this *new* way of looking at feminine hygiene with her good friend Hildegard, an elderly woman who survived the second world war before emigrating to Canada as a young married woman. Hildegard looked a little amused as she explained that the idea is not new at all. There was a time when there were no menstrual products available as no one had invented them yet. Instead, women made do with what they had. In some areas, moss and soft grasses were wrapped in soft cloth or strips of soft hide. In Hildegard's case, her family sewed their own pads out of layers of cloth—usually old clothing since during wartime everything was scarce. Even today, regions on our earth have never seen a sanitary product.

Lillian realized that she was wasting several pads each cycle due to anticipation that her period would arrive early or last longer. She decided to try a reusable pad product that would at least eliminate this waste. Within a short time, she was using them for her entire cycle. Two years later the reusable menstrual pads that she purchased are still going strong.

Anxiety over handling the used products is understandable, but unfounded. Lillian reuses a one-gallon ice cream bucket with a tight fitting lid and fills it half full with cold water. This bucket is kept near the toilet to store the used pads in. A couple times a day, she pours the water down the toilet and adds fresh water. Every couple of days, she will run them through the laundry machine. The procedure is simple and she finds it does not create any unpleasant odors. Disposable diapers are handled in much the same manner.

Within the first few cycles, Lillian realized she was no longer experiencing the skin irritation caused by disposable products. As an added bonus, what would have been spent on disposable products in one year paid for all of the cloth pads. Since then, we estimate saving $100 annually. In a lifetime, a woman can spend a lot of money on disposable menstrual products. Reusable sanitary products are ideal for traveling to other countries where these products may not be readily available.

Benefits
• Support non-health-threatening alternatives and reduce strain on the medical system.
• Reduce waste and keep toxins out of the environment.
• Experience more comfort.
• Save money.
• Travel without worry.
• Send clear messages to disposable product manufacturers.

ORGANIC GARDENING

There is no better way for a gardener to grow than organically. You avoid all the toxins and carcinogens found in many of the commercial fertilizers and insecticides. Not only is the nutritional value retained, but also it is increased by up to 3 times as much as commercially grown crops. Chemical fertilizers also harm vital microorganisms resulting in an unbalanced soil and weakened plants.

Waging a chemical war on pests is a losing battle, as insects and weeds have proven an ability to develop immunities to pesticides, herbicides and fungicides (collectively known as biocides). Biocides are turning our backyards into such toxic sites that they have become uninhabitable for delicate beneficial insects and wildlife - yet we encourage our children to play in them! Biocides, toxic to bees whose population in Canada is steadily declining, are a threat to natural pollination of crops and indigenous plants. When used to control weeds, biocides contaminate ground water and neighboring springs and wells.

Inter-planting crops with both companion and beneficial plants reduces pest problems and improves growing conditions while attracting pollinating insects. Intensive growing allows little room for weeds to flourish. These methods of gardening avoid biocide use while increasing yield, vigor, and even the flavor of the crop.

Growing Non-hybrid Crops
Hybrid crops, altered genetically, often incorporate genes from

the animal kingdom and other unnatural substances, such as vaccines and biocides. Typically, hybrids are created for uniform appearance, shipping ease, and resistance to biocides. Their mutated or sterile seeds are unlikely to germinate or produce true to form, making seed-saving impractical.

Choosing to grow non-hybrid varieties is an effective way to support and even improve the genetic diversity available in North America. There are many terms applied to non-hybrid crops. *Heirloom* (grown for 100 years) and *heritage* (grown for 500 years or more) are *open-pollinated* crops. Open-pollinated means we leave the pollination duties to the insects, allowing the plants to produce seeds naturally without manipulation. Furthermore, non-hybrid strains offer a much larger variety of colors and shapes than is commercially available.

Hybridization occurs naturally when wind and insects carry pollen from hybrid garden plants to wild, indigenous plants disrupting nature's delicate balance. Mutations and loss of diversity in our wilderness areas is a very real danger. We can prevent this from happening by choosing to grow non-hybrid varieties.

Benefits
- Avoid toxins and carcinogens and retain nutritional value of food.
- Prevent ground water contamination.
- Reduce pest problems.
- Improve growing conditions.
- Support genetic diversity.
- Decrease overall waste and cost.

ORGANIC WASTE

Food waste increased by 1.2 million tons in the 25-year period between 1970 and 1995, only an estimated 4.1% of that was recovered. Such waste increases have prompted areas, such as the province of Nova Scotia, to refuse food waste at all their landfills.

Soup Stock

Ever wonder how your grandmother was able to make such fantastic soups? Here, we will share how to convert kitchen waste into a nutrient rich broth.

The process is really very simple. Save any clean, disease free vegetable debris, as well as any herb branches, and reserved cooking water from pasta or corn, for instance. Clean eggshells were once commonly used in stocks as a source of calcium. If you wish to try it, old-timers recommend adding a splash of vinegar, which will leach calcium out of the shells easier.

Keep a recycled margarine-type container or a bread bag to store all the vegetable debris and any liquids inside the fridge for a few days, until there is enough to warrant running a pot on the stove. When there is enough to brew, simply dump the containers' contents into an appropriately sized pot and add enough water to cover. Bring to a boil, cover, and turn off the heat. By the time the pot has cooled, most of the nutrients and flavors will have cooked into the liquid. If you have added leftover meat or bones, let it simmer, covered, for at least 20 minutes (more if the meat is raw) to get more flavor and nutrients in your broth.

Place a fine sieve over a large bowl, preferably one with a pouring spout, and strain the broth. The cooked mush that is left will break down quickly in the compost or worm bin. If you used meat or bones in your stock, seal all the debris in a plastic bag before dumping in the garbage. This will help isolate unwanted odors.

Pour the cooled broth into a clean, recycled juice or milk jug for storage. If you plan on freezing these jugs, be sure to leave a few inches to allow for expansion.

We have so many uses for stock in our house besides just soups and stews that we typically can't keep up with the demand. We use it in place of oil in our stir-fry dishes, to cook beans and grains in, and we even use it in some of our breads. We were surprised at the difference in flavor homemade stock brings to meals. Occasionally, we add a bit of bulk soup stock powder if the stock is a little bland.

By doing this one step, we not only reduced the volume of our waste and the need to buy packaged broth products, we also extended the value of our shopping dollars and improved our nutrition. In addition, we reused containers formally destined for the landfill.

Coffee Grounds

Those messy, often smelly coffee grounds can quickly urge us to take out the kitchen garbage long before the bag is actually full. It does not have to be this way though.

Coffee grounds are an excellent addition to the compost or worm bin. Paper filters can be put in your compost as well, but they take quite a while to break down and may contain dioxins due to bleaching processes—not exactly something we want in our coffee, let alone in our garden! Reusable coffee filters, available at most grocery stores, are really very simple to use. Hold the coffee basket upside down and lightly tap on the bottom until all the grounds come out. Rinse and make the next pot or place the basket in the dishwasher. We did some comparing and found the cheapest prices for the following calculations. 1000 disposable paper filters—just above the average home's annual paper filter consumption, costs around $14. We have bought a reusable filter for $9 about 10 years ago. The filter is still going strong with no signs of wear saving us about $125 so far.

Coffee grounds can be used as a 1/2" thick mulch around any acid-loving perennial bed (i.e. strawberry), but be sure to leave some room around the base of the plant and keep leaves clear of the grounds, or you may experience some burning due to the high nitrogen content. Incidentally, because of it's high acidity level, neither ants nor slugs like this mulch. When we heard this, we tried it on one home with an ant problem. Lining the exterior with a 3" X 1/2" coffee ground mulch, we slowly worked our way around the home. It seemed we were chasing the ants around the house as we went. A year after this ring was placed, we noticed some ants coming back—so it may be wise to make the coffee ring an annual chore.

Drying coffee grounds may seem a little fanatical, but it allows for storage until a soil amendment application is needed. First, spread the coffee grounds on a reused foil tray or an old baking dish. Place in a preheating oven and when ready to bake in the oven, simply remove the tray and place it on the burner that the oven exhausts through. After baking, turn the oven off, replace the tray, close the door and allow the oven to cool. All in all, this process takes about 20 minutes, depending on how wet the grounds are and how warm the oven is. Alternatively, you can set the tray on top of your water heater, near a wood stove, or on your fridge—reusing their heat. Homes that use forced-air-heating systems can place the tray on a rack over the floor vents.

Reusing a clean container with a good fitting lid, bend the foil tray slightly to attain a pouring edge and then dump the cooled grounds in the container. Leave the lid off or at least slightly ajar for a day, to allow for any further drying. Be sure to label the filled container clearly and store with the garden supplies.

Although coffee grounds contain 4% Nitrogen, 1% Phosphorus, and 3% Potassium along with trace minerals, they contain too little phosphorus and calcium to be considered a *complete* fertilizer. This is why we add equal amounts of ground eggshells and dried banana peels to our mixture before using as an amendment to our soil.

For seedlings, mix up to 1 cup into each gallon of soil, depending on your soil's pH level; grounds can burn young transplants if too much is used. Before transplanting to the garden, place 1/2 cup in the bottom of each hole and cover with a 1" layer of soil.

Feed trees and bushes every month until first frost by scattering a 1/2" layer of this mix around the base of each plant. Then lightly rake it in with a hand rake or with a large garden rake, depending on the size of the area.

Black Tea Bags

Due to the tannic acid found in black tea, bags have been used to ease swelling and tiredness for eyes. The cool, used bags are placed directly on the eyelids for about 10 minutes. They are also an excellent soil amendment because they are rich in nitrogen and some trace elements.

Cut along the two adjoining sides of the tea bag to ease the extraction of the grounds. Either contribute directly into the compost, worm bin, or garden soil, or dry and store for use as a soil amendment (like coffee grounds).

Eggshells

Like all organic matter, eggshells can be contributed to the compost and worm bins. Extremely rich in calcium, eggshells also make an excellent soil amendment. They are especially useful for members of the Brassica family (broccoli, cabbage, chard, kale, etc.). Calcium-greedy crops benefit from 1/2 cup of shells placed in their transplant hole. On our family's organic farm, we supplemented the chicken's diet with dried ground shells. Applied as mulch around slug-prone areas, the sharp edges of the crushed eggshells deter the creatures from returning.

To dry eggshells, first rinse clean and place on a reused foil tray. Place in the oven either during preheat or cool-down times to utilize this energy, or on its lowest setting. They will also dry in just a few minutes on top of the wood stove. It is important to note that eggshells burn easily. Lightly crush either by hand or in a blender (not to a powder) and store in a reused container with a good fitting lid. Large peanut butter or mayonnaise containers are good for this. Leave the lid slightly ajar for a day allowing for any further drying. Label containers clearly and store with your garden supplies.

Alternatively, use the eggshells as biodegradable pots for your seedlings. This works very well for plants that do not need to be very big before transplanting. However, it works best if the shell is cracked before planting to allow the roots faster access to the soil. To make a complete soil amendment, see *Coffee Grounds*.

Banana Peels

Banana peels, rich in Potassium (essential for plant growth and disease resistance), also make a good soil amendment. Be sure to chop as finely as you can before drying in a warm airy place—on the fridge, over a heating vent, or on top of the water heater.

Store in reused containers with good fitting lids, leaving the lids ajar for a day or two, allowing for any further drying. Label the containers and store with garden supplies. To make a complete soil amendment, see *Coffee Grounds*.

Autumn Leaves

At this time of year, we begin to see leaves flutter to the ground, turning green lots into a carpet of color—and later, slime! But what valuable slime it is!

Leaves break down quickly and are excellent for amending garden soil. Mainly cellulose-based leaves are packed with calcium, magnesium, nitrogen, phosphorus, and potassium. They also create the much needed loam (organic matter that is responsible for retaining water) in the soil. Neighbors, lawn maintenance companies, schools, community centers, and motels are all good sources of free leaves.

We like to run our lawn mower, which has a grass-catching bag, over the leaves—thus, chopping them and picking them up all at once. We then build a pile about 2' high by alternating 4" of leaves and thin layers of soil; soil contains the microorganisms needed to break down the compost. If you happen to have fresh cut grass, layer it in this pile. Grass is an excellent source of nitrogen, creates heat, and releases moisture, which breaks down the compost very quickly.

Our pile has fully composted and produced ready-to-use humus in as little as 6 weeks. Keep the pile moist, but not too wet. For maintenance, the most important tool is your compost fork. Use it to turn the pile at least once every week. Turning the pile may look like a lot of work but it is actually quite a light and fluffy chore. However, if you cannot turn your pile, no problem! Simply let the pile sit until springtime; it should be ready to use by then.

Lazy Black Gold

Composting automatically reduces the average family's wastes by 30% and produces a fantastic soil amendment. Bagged humus (broken down compost) is quite costly, so reusing organic waste will actually save the household money. Besides reusing waste formally destined to the landfill, composting families are also doing their bit against global warming. When organic waste is covered in the landfill, it is deprived of oxygen, which creates an *anaerobic environment* that reacts with other waste materials, producing leachate and methane—a threat to ground water and a producer of greenhouse gasses.

It is also interesting to note that composts have been found to generate energy, which can be tapped and converted into electricity using copper wire. Remember the old science class project of placing a copper rod and a zinc rod into a potato? This simple battery can produce enough electricity to power a clock for 2 weeks! Imagine what a compost pile could power.

Composts need not be labor intensive or rushed projects. However, the more the compost is turned, the more quickly it will break down. We have had compost break down in just a few weeks with diligent turning. Yet, piles of layered matter sit without being turned once and still break down. And they do not have to be considered unsightly or wasted spaces. Squash and gourd plants thrive in these compost heaps and improve the appearance. They can also be camouflaged with sunflower, sun-root, amaranth, or legume plantings. Lillian saw a magazine photo of a compost bin that had a healthy crop of peas growing up its sides, completely camouflaging the pile from all angles. We experienced a sensational success with fragrant sweet peas, violas, snow cloth alyssum, and a couple of poppies around one of our bins last year.

Many efforts have been made using several types of bin designs and compost techniques to speed the conversion time from compost to humus. A compost can be as simple as an unsheltered pile of organic material left to compost on its own, or it can be made into a time consuming, and even costly project.

Alternate layers of green material (grass clippings and kitchen waste) and brown material (aged manure, soil, leaves, paper towels, napkins, tissue paper) constitutes a compost. Shredding materials before contributing to the pile is helpful in reducing decomposition time, but not necessary. Keep the pile slightly moist, but not wet; it will aid the microorganisms in the pile. There are microbiological mixes commercially available to help accelerate composting. Alternatively, the addition of comfrey or nettle leaves will also speed up the process. Regardless of the type of ingredient, each should only be placed in 4" layers alternating with 1" layers of soil. It is important to note that manure contributes to odor, pests, and disease— therefore use thinner layers.

Any organic matter can be composted, but animal waste can attract scavengers and insects, and may harbor transmittable diseases. Although manure will increase heat in the compost pile, it also increases the salt content in soil and may contain undigested seeds that will sprout as weeds in the garden; these need at least 2 years to fully decompose. Both dog and cat wastes should not be used in food crop areas because of the risk of disease. However, this humus would be fine for trees, flower gardens, and bushes.

In the past, we have pinned landscape fabric over bins with logs or lumber for several reasons. It is black and attracts heat, while at the same time it is porous, allowing for proper ventilation. Covering the compost reduces insects and scavengers, but most importantly, maintains moisture content and reduces the germination of unwanted weeds. Anything like straw, soil, or boards will do. It is important not to seal the top of the compost off completely. The compost should not smell bad, although it can smell like rotten eggs when oxygen starved. Nitrogen rich composts are slimy and smell like ammonia. If any smells do occur, turn in some leaves or straw, and cover with soil.

To get additional beneficial nutrients and trace minerals, some humus-crazed folks, like us, go hunting beyond their own yard and kitchen for compost ingredients. This involves picking up bags of leaves and grass from friends with large yards, who are only too happy to find a use for their yard waste. This can greatly increase the bulk of your compost pile, while reducing the landfill contributions of the community.

Sources of Organic Waste to consider:
• Coffee grounds—local coffee shops and staff coffee rooms.
• Used barley grain and used Hops—breweries.
• Leaves and grass clippings—parks dept, landscaping or lawn-maintenance contractors.
• Seaweed—raked beaches.
• Fruit pulp—juicers and wineries.
• Hay—farmers * Note: Compost hay for 2 years to kill seeds. There may be seed-free hay available from some farms.
• Kitchen waste—bakeries, cafes, or restaurants.

Benefits
• Composting alone automatically reduces the average family's wastes by about 30%.
• Reduce the need to buy packaged broth products, while extending the value of your shopping dollar and improving nutrition.
• Save money in soil amendments, mulches, pest deterrents, biocides, stocks, and plant pots.
• Reuse waste and containers formally destined for the landfill, while making alternative uses of the heat sources in the home.
• Save $14 annually by employing reusable coffee filters.

VERMICULTURE

Vermiculture, in use since before the 1950's, is the act of employing worms to convert organic residues into a valuable resource. Worms ingest microorganisms (fungi, protozoa, algae, nematodes, bacteria) and organic matter, then excrete castings. Also known as Vermicompost, castings contain a wealth of nutrients and microorganisms. When added to soil, it increases the soil's health tremendously. Today, thousands of companies and cities make use of vermiculture because it is profitable to reduce waste headed for the landfill.

Why would you want to handle worms when compost could be taken outside easily enough? Odorless, clean, and inexpensive worms are ideal for those who generate smaller quantities of waste. Vermiculture can be done year-round and is faster, cleaner, and has fewer pests than traditional outdoor compost heaps. Indoor worm bins—protected from wildlife, avoid habitualizing the animals to eating human garbage. In the garden, worm pits can help process large amounts of waste, such as autumn leaves or grass clippings. Once created, these compost piles are unattended and will break down faster than normal with the help of worms.

Children seem fascinated by worms and parents tell us that their worm bin is not only an excellent education tool, but the worms have even become family pets!

There are approximately 4400 known varieties of worms, but for vermiculture purposes, redworms (Eisenia Foetida and Lumbicus Rubellus) are the most commonly used. Redworms burrow horizontally, do not create extensive homes, and reproduce cocoons

2-3 times a week—which quickly mature into productive worms. Because of this, red worms can handle bed disruption. Night Crawlers, on the other hand, have extensive vertical tunnels and do poorly in soil that is disturbed. Night Crawlers produce fewer, but larger, cocoons, which are slow to mature.

Cool locations are best for worms, which will die from exposure to extremes of heat or cold. Any place that has temperatures ranging between 65-75°F will do. Cupboards, closets, basements, garages, and sheds make ideal sites for a worm bin. If temperatures drop too low, there are some things you can do to help your little friends. Insulate bins with 2" Styrofoam. Set a bird bath heater to 40°, immerse in a 2 gallon container of water, and place this in the center of the worm bin. This will keep the bin just warm enough for the worms to congregate and survive. However, do not despair if you forget and leave the worms exposed to poor weather. The worms may die and decompose in the bed, but their cocoons will still hatch in the spring.

To determine how many worms your family needs, simply weigh one average week's worth of the household waste, not including excess garden or yard waste. For each pound of waste, you need 2 pounds of worms. Typically, a family of 4 requires about a 1' x 2 'x 3' sized container, which can be anything from a three-tier plastic bin set up, to a homemade wooden box. The container should have 1/4" holes drilled 1/2" apart for drainage. Applying legs or propping the bin with bricks or boards, and setting over a drip tray will ease the cleaning chore and ensure proper aeration and drainage. If there is a need to move the bin occasionally, consider mounting it on a set of wheels.

Stacking worm bins works quite well and is the method we use. As the worms finish the first bin, they will work their way up through the holes on the bottom of the second bin, where there is fresh bedding and food waiting for them. This is a great way to separate the worms from the castings without even having to touch them. Otherwise, harvesting requires emptying the entire contents on a large surface and using lights to chase the worms to the bottom of the pile. The castings are then brushed aside from the retreating worms.

Before your worms go in their new home, it must be prepared properly. The first layer consists of bedding (1/2" strips of newspaper or cardboard, sawdust, leaves, seaweed, or dry grass clippings). Fill the box 3/4 full with layers of organic compost materials with alternate 1/4" layers of sand, crushed eggshells, or soil. Worms' bodies are long digestion tracts without teeth, so they require the grit to digest food. Add enough water to make a moist, but not wet, home and set the worms on the top. Though worms cannot see, they are light sensitive—a cover for the bin is recommended. This will reduce water loss, prevent flies and maintain darkness for the worms.

Periodically mist the bin so that it does not dry out. The ideal moisture level for a worm bin is about 65% humidity. Worms breathe oxygen through their moist skin and move via hydraulic pressure, so a moist, well-ventilated environment is essential. Bad odors signal there is too much food or water for the worms to handle the bin properly. Feed the worms finely chopped kitchen waste once a week, as they do not like to be disturbed more often than that.

Hermaphroditic worms mate end to end resulting in both parties being impregnated. In ideal conditions, a worm can live 4 years and produce up to 3,000 offspring a year. However, the populations are self-regulating—as conditions get crowded, the worms will breed less frequently and mortality rates will stabilize their numbers. Worms are very considerate; when their babies have hatched, the adults will leave the main food source for them. They will actually move away to previous feeding grounds, even if they end up perishing from lack of food.

When a bin is harvested for castings, we release any excess worms into the garden beds and composts. Very industrious creatures, they break up compacted earth by their tunnels, allowing for better circulation of water and air in the soil.

One handful of castings in transplant holes eliminates the need for fertilizers and results in beautiful, pest free plants. It is important to note that castings like all manure, increase the salt content of the soil and should be used sparingly.

Outdoors, use worm pits to manage large amounts of waste, such as autumn leaves or grass clippings. These are simply unattended compost piles; however, the alternate layers of soil, sand, and organic matter are piled in a heap. In the center of the heap is a hole (as if a pole was pounded into the center then removed) where daily waste is contributed. Turn the pile occasionally to improve air circulation and avoid compaction. Damp burlap or cardboard over the pile will help keep it cool and moist in hot weather. Like worm bins, the worm pit should not be allowed to become too dry.

If you find you do not produce enough waste for your worm pit, search within your community for potential sources. For some ideas, see *Organic Waste*.

Benefits
• Increase the amount of microorganisms, nutrients, and minerals in the soil with worm castings.
• Composting alternative for those who generate small quantities of waste.
• Eliminate exposure to winter chills and protect from outdoor wildlife.
• Worms fascinate and educate children.
• Worm tunnels allow for better circulation of water and air in the soil.

WATER USE

Of the world's water, 97% percent is salt water, 2% is locked in polar ice and underground springs, and only 1% of it is actually available to us.

25% of the world's fresh water supply lies within Canadian borders, yet our chances of facing drought increase as North American aquifers continue to shrink.

In 1992, the average Canadian used 150 gallons every day. In comparison, just running the tap for 60 seconds is all that some people in developing countries have to survive on for 24 hours!

Recently, *World Water Day* celebrants heard the United Nations warn that over 2.5 billion people will face severe water shortages by 2025 if the world continues its present water consumption rates. Already, 1.1 billion people do not have access to safe water.

Household water use accounts for 5-10% of the total used worldwide; most of that is used here in North America. There are many things that you as an individual can do to better manage this precious resource.

The average North American sink is responsible for using 8 gallons of water daily. Running the tap while brushing your teeth can waste 2-3 gallons of water each time, so turn it off until you need to rinse.

Inspecting the household water system for leaks is a powerful waste reduction step. Did you know that the leak from one worn washer could fill a swimming pool in a year? Now consider the cost if that is hot water. One drop per second can equate to gallons of water every day. Now, that is monetary encouragement for fixing leaky faucets!

Bathrooms

According to *Environment Canada*, 65% of household water use is in the bathroom. Each of us flushes 9,000 gallons annually, or 25 gallons a day. Traditional toilets use about 5 gallons of water per flush. Installing water-conserving toilets saves the average household 32,000 gallons of water annually. For those who do not have this option, reusing a 16-ounce bottle is an alternative. Filled with water and placed clear of inner workings, the bottle displaces the water level, greatly reducing water use per flush. For the mechanically adventurous, upgrading the inner workings entirely is also an option. According to CBC radio, there are areas where waterless urinals are successfully employed in public washrooms, resulting in 40,000 gallons of water saved per year.

Toilets are not the only water wasters in the bathroom. A shower, operated 20 minutes a day for one week, uses 700 gallons of water, whereas a single bath alone can use as much as 34 gallons of water. Showers also tend to consume 40% less hot water than baths. So choosing to take showers instead of baths saves energy and water. This is something many of us are doing already because showers are so much faster than baths. Baths are now seen as a leisure activity, used occasionally to de-stress one's self or to soothe aching muscles.

Showers can be even more efficient when low-flow showerheads and faucets are installed. Shortening the length of time in the shower is another step to reducing water consumption; the most efficient shower is less than 10 minutes. On our family farm, we once had an outdoor solar-heated shower that we turned off while scrubbing and shaving. Some water conservationists practice this idea regularly in their own bathrooms. Better yet, try installing a showerhead regulator. This is a pipe extension with a tap built into it. The advantage here is that you are able to interrupt the water's flow while maintaining the temperature.

To go to this extreme is not easy for everyone to do. However, if every North American did these things, it could reduce household water consumption by a staggering 75%.

Kitchen

Rather than running the tap when cleaning vegetables, use a bowl of water. Later, reuse it to water outdoor plants. Reusing water from rinsing out the coffeepot for outdoor plants is something we do all the time. Rich in nitrogen as well as some trace minerals, it should be diluted further before using.

After meals, scrape your dishes into the compost bucket before rinsing. While rinsing, place other soiled dishes and utensils underneath while you work; it will begin the presoaking process. If you have cooked a meatless meal, this rinse water can also be used for outdoor watering.

Save about 5 gallons of water per washing by doing dishes in a few inches of hot soapy water. It may seem funny to do this, but by rinsing with hot tap water into the wash sink, the level will slowly increase and will maintain a piping-hot temperature.

The average North American home consumes 7 gallons of water a day for the treatment of drinking water alone. It is sad, but *pure* water is not so readily found anymore, yet it is very important for our health. In our youth, it was nothing to drink from a mountain stream while out hiking. Nowadays you had better pack in your own good water or a purifier unless you want to risk getting ill.

You can reduce the amount of pure water wasted with some simple water tricks. Keep a jug of water inside the fridge, which reduces the need for ice cubes. Use purified water for cooking and drinking only, and tap water for your other needs. Do not forget to water your plants with any remaining liquid in your glasses. Dumping that purified water down the sink just does not make *cents*.

Laundry Room

How often have we been reminded·to use proper settings on the laundry machine? It seems there is good reason for those reminders.

Appliances are responsible for 21 gallons of water use daily by the average home. A clothes washer alone uses 7-15 gallons per cycle. This is why it is important to use proper size settings. For similar reasons, be sure to completely fill your dishwasher before operating.

Lawns and Gardens

Choosing heat and drought tolerant plants for landscaping will reduce your outdoor watering needs. In the garden, exposed soil is quickly robbed of moisture by the sun and wind. Covering naked soil with mulch prevents this, with the additional benefit of controlling weeds. Watering gardens and landscaping with a drip or soaker hose consumes only 20% of what overhead methods use.

There is great dispute about whether lawns are a good idea or not. An average lawn is responsible for consuming 21 gallons of water daily. Three square yards of grass annually remove 1 pound of airborne particles and will produce enough oxygen to supply one person's annual oxygen intake. It has long been known that golf courses improve humidity levels for the surrounding community. Undoubtedly, there are many things to consider before taking a side in this debate.

For those who have the choice of putting in a new lawn, there are many seed alternatives that can fit your particular needs. Some grasses are good for shade, others for high traffic zones. Some grass varieties only grow to 4" tall. Some lawn mixes combine grasses with wildflowers. There are many alternatives for green cover to choose from, such as creeping thyme.

By mowing lawns in the early evening and maintaining a height of at least 2", watering needs are greatly reduced. Water lawns in early morning or evening hours when evaporation is less of a concern. Be sure to use the appropriate water setting for the area so that you do not water the sidewalk, a building, or the street. Some types of landscaping require frequent, but light watering. Others require deep and less frequent watering. Be sure to know what your plants need— a nursery should be able to help you with this. Using a timer on sprinklers is a great idea, but automatic underground sprinkling systems are probably the most efficient, effortless method available.

Consider installing a *Gray Water* recycling system. A search on the Net or the library will provide you with information on this subject. The term *Gray* is based on the color of water gathered from sink, tub and laundering areas due to soap residues. This water is filtered, then treated with various water plants in a series of ponds and fed to outdoor landscaping. The now infamous Body Shop has a gray water system in their processing plant that cleans the water so well it can be used for drinking. Nature's own way of treating water through the use of wetlands is the model used here.

Many people collect rainwater for their individual landscaping needs. This is easily done by simply placing a barrel where it can catch the rainwater—such as under a downspout from the gutters. Rainwater is nutrient rich, so valuable to some that they have made elaborate collection operations. One Texan gardener we read about watered her entire garden by rainwater alone—and they do not get very much rain there!

One local gardener supplements his rain-collection barrel with tap water using toilet tank parts to regulate the water level. The barrel has a hole at the base where water continually seeps out to his garden hoses. In our garden, Dave has it so nicely set up that there are control valves at the end of each bed for better control to meet the particular needs of each crop. These are all excellent methods that allow for more free time as well as better water use.

Using rain gardens for city storm water management is a movement that is rapidly growing. Runoff from driveways, parking areas, sidewalks and streets is laden with pollutants. Allowing this to flow directly into natural waterways untreated seems like a crime. Many communities are incorporating rain gardens to combat this problem. Usually located in runoff collection sites and flood-prone areas, they are filled with plants with a proven ability to clean polluted water. The additional benefit of such gardens, besides the increased greenery, is the wildlife it attracts. Many of the plants produce flowers and fruits that sustain hungry birds and butterflies.

Because rainwater has no minerals, one can use it in place of demineralized water for clothes irons. Rainwater is also useful for washing vehicles. Speaking of which, did you know that washing your vehicle at home with a bucket of water and then hosing it off is the most ecologically friendly way? If you avoid having a running hose and instead use a spray gun when washing the car, it will save up to 150 gallons of water each time. Keeping your vehicle clean, especially during cold seasons, can extend its life, but during less dirty seasons, sparkling clean vehicles are not as necessary. Choosing to wash the car a little less often can save a huge amount of water. For instance, if you hand-wash your car every 7 days and choose instead to wash every 8 days, you have saved between 60 and 120 gallons annually. And do not forget to hand wash vehicles on the lawn, where the water will do some good rather than going down the street drain.

Swimming Pools

Pool covers reduce evaporation by up to 95%; without a cover, a pool can lose 1 inch per week in the peak of summer. In addition, leaks in the pool can cause algae and other water problems, therefore it is important to detect even the smallest leaks.

By incorporating any one of these ideas, you will have contributed to proper water management. Many communities, partially due to overpopulation and environment, are now forced to install residential and commercial water meters and charge rates based on use. For example: Just 10 years ago in Kelowna, we had never heard of water meters. In the late 1990's, residents were asked to water at certain times and days to reduce the water demand. By 2002, virtually every building had meters installed. And so, for those who reside in metered communities, water conservation will have much more meaning.

Benefits
• Fixing leaky faucets results in tremendous energy and water savings.

- Installing water-conserving toilets saves gallons of water daily.
- Take showers instead of baths and save both energy and water.
- Save water and money spent on fertilizers by using rinse water from coffeepots.
- Save 80% of your gardens and landscaping water consumption.
- Hassle free watering with automatic underground sprinkling systems.
- Save gallons of water each time you wash the vehicle.
- Replace demineralized water with rainwater and save money.

PART THREE

TREES

When asked by a reporter what the most effective thing is that an individual could do for the planet if they had only one day, Martin Luther King's response was, "plant a tree." There is ample truth to back up this statement. Presently, plant life absorbs roughly one half of our fossil fuel emissions. Air pollution is a common complaint in urban areas and because trees are effective at carbon sequestering, especially hardwoods, it makes sense for us to plant many more. Often called *lungs of the Earth*, trees also convert carbon dioxide into oxygen making their presence utterly vital to our existence. Recent studies show that evergreen trees, especially fir and cedar, give off therapeutic fragrances. A nine-year study conducted by the University of Delaware discovered that hospital patients who could see trees from their window actually went home earlier. Trees beautify the land and provide privacy, shade, and wind protection. Providing a habitat for wildlife, they also screen out noise and maintain moisture in the soil.

Establishments and homeowners save a lot of money in energy costs, simply by planting trees. A properly placed tree can reduce the temperature of your yard and home by 3-7°F due to the shade and evaporative cooling qualities. For protection from wind and snow, coniferous species work best when located on the prevailing wind side of the buildings. Plant deciduous trees on the south side to provide shade in the summer, but not in the winter when the leaves are off the tree. For additional incentive to the homeowner, landscaping can increase real estate values by more than 10%. Trees break up air patterns, causing soft breezes. This coupled with the moisture released from the leaves and shade, combat the *bake effect* found in urban areas. The bake effect is the result of heat absorption (concrete and asphalt) combined with reflective surfaces (glass and steel).

Victoria School in Saskatoon started several in-house programs for the children to participate in. They offered tree seedlings for the kids to plant at home or near unstable areas such as riverbanks. They have also hosted clean up campaigns for local green areas. Others have begun urban tree-starting garden projects. Tree plots can be surprisingly small and yield between 30 and 50 seedlings a year. Some, like the University of Guelph's Arboretum in Ontario, are growing indigenous trees from seed collected from within 6 miles. These efforts benefit everyone by bringing nature into our communities. Feel the refreshing coolness of a well-vegetated area on a hot, muggy day and you will see what we mean.

It is estimated that a single 50-year-old tree can produce $164,250 of environmental value between oxygen production, air pollution control, water protection, and prevention of erosion. As they are one of the most called upon resources in our day, the preservation of trees and forests should be top priority in the plight to preserve the environment. Although recycling reduces the contribution of paper products to the landfill by at least 30%, it is equally as important to reduce and reuse this resource as much as possible before recycling.

Reduce

Save trees by choosing to use computer disks and emails for data storage and communications instead of printed versions. Paying bills online and utilizing automatic renewal options available for magazine subscriptions reduces both paper use and postage costs. Consumers can affect the ways we use our forests by purchasing eco-certified paper products and paper containing recycled fibers or alternative papers.

The first people to make paper used bark, skin, reeds, or straw. Europeans used to make paper out of rags, until there was a shortage of rags. With this in mind, it is ironic that modern people think of tree-free paper as *new-age*. Papers made from kenaf, sugar beet, corn, algae, sugar cane, grasses, and bamboo are not unheard of on today's market. Increasingly, we are seeing paper produced with recycled (a.k.a. post consumer) and alternative fiber content.

Farms of fast growing fiber crops such as willow trees may help alleviate the pressure caused by our dependency on nonrenewable fossil fuels and the declination of natural forests. Willow trees are a good source of biomass fuel and of cellulose—used for making paper. Sweden is testing over 11,000 hectares of this crop, which is used to fuel regional hospitals and schools. Willow plantations require less maintenance, are ready to cut within 2 to 3 years, and will regenerate from the root after each harvest. In the 21-year life span of a willow, it will produce 10 times that of a poplar or aspen, and 100 times that of a coniferous tree.

Hemp plantations, used for making rope, boards, cloth, nutritional and beauty products, and paper, produce twice the amount of fiber than cotton. Hemp fiberboard is stronger than similar wood products. The crop grows well in global warming conditions and counteracts it by its massive carbon sequestering ability. It is theorized that hemp crops actually extract unwanted toxins out of the soil. The potential for hemp is huge and many countries, including Canada and parts of the United States, are growing commercial crops.

Reusing lumber

If you have building skills and are equipped with tools, consider reusing lumber for projects around your home. Children's playgrounds, bat houses, bird houses, compost and worm bins, garden beds, cold frames, and stairs for entrance doors have all been successfully made by reusing lumber that was considered waste. Over the years, Dave has constructed greenhouses and workshops entirely from old pallets, saving thousands of dollars in material costs. Railway ties and other treated wood, while not appropriate for edible crop areas, are ideal for landscape terracing, steps and pathways.

Many companies are jumping to meet the demands of earth-conscious consumers. T-rex Demolition in Vermont was developed to fill the need for recycled building materials and has reached an incredible recycling rate of 80%. They send tons of steel and iron to scrap yards and glass to pavement producers. Any wood materials are reused as lumber, burned for fuel, or converted into paper. Renovator's Resource in Halifax similarly scours renovation and demolition sites for useful or interesting materials to be sold to builders and contractors. They produce their own line of furniture made from recycled materials and are specialists in the relocation of old buildings.

The Journey of Recycling Paper Products

Nearly one half of the harvested trees in North America are used in the manufacturing of the four main categories of the paper industry. Corrugated paper or paperboard accounts for a little more than 50% of all paper produced, high-grade printing and writing paper reaches 30%, newsprint accounts for 8%, and tube products approach 7%.

We could learn from our European neighbors. More than 75% of Dutch paper needs are satisfied by recycled paper products. Recycling paper products can go a long way in saving the environment and aiding the health of Earth's inhabitants. The paper recycling industry creates 75% less air pollution, 35% less water pollution and uses 60% less water than that of virgin materials.

Recycled paper is formed into a pulp and inks are removed with detergents. The pulp is then filtered and pressed. Recycled paper content can be found in new paper products, in insulation, animal bedding, and cat litter products. As you can see, there are plenty of positive reasons to continue recycling paper.

Tourism

Tourism is one of Canada's many assets, and we need people to keep returning to our beautiful country. Greenery and lush surroundings are major reasons for tourists to return to an area. North

American communities are working together to create patches of greenery to beautify and cool their towns, preserve wildlife, and provide other environmental benefits. Because both looking at and working in gardens are known therapeutic activities, gardens are used by hospitals and old-age homes. Cities make full use of their gardening dollar by also using the areas as educational tools for schools and citizens alike. Gardens are used as cropland for the poor or for those who do not have access to land, and as tourist attraction areas. Cities and towns can be famous simply for their piece of greenery. Consider Vancouver's Stanley Park and Victoria's Beauchart Gardens. Let us not forget that tourist-friendly areas create jobs and residual revenue for the community.

ENVELOPES

Extending the life of any paper product by reusing can help lessen the consumption of our forests. Homes and offices can save money through reusing envelopes, not only in actual cost, but also in storage and handling. Major corporations are now starting to practice envelope reuse. Federal Express, for example, has recently employed a reusable envelope they send to millions of their customers. Instructions are included on how to refold the envelope for their payment. Some two-way envelopes even use available space for a letter, questionnaire, or a change of address form.

Reusing envelopes before recycling reduces consumerism. Less money is spent and less packaging is sent to the landfill. If you open envelopes carefully, at one end only, you will be able to use them again. Manila envelopes can be adapted to store important documents, certificates, and diplomas in your file drawers—keeping them safe from harm and separated for easy access. Various sized envelopes are ideal in a file cabinet to store receipts, tax information, or bills. The smaller ones work well for storing garden seeds. Seal by folding the opening over twice and secure by taping or stapling. You can write the name of the seed on the envelope.

Reuse envelopes for mailing purposes - especially those extra ones often included with magazine or credit card memberships. Simply place a label over the printed address and fill out as usual. One option is simply to secure an addressed strip of paper with tape. Alternatively, cut an unused videotape label in half, placing it over the previous address. CD and cassette labels work well, but you may need to use more than one. Now write the address on the label seal the envelope and mail away.

You can also create custom labels on your computer printer. Some have gone as far as including the message, "we reused this envelope to do our part for the environment. Won't you do the same?" This is a very effective way to promote reusing, as many an eye falls on that envelope during its journey.

There are quite a few other ways to reuse envelopes, limited only by your imagination. Cut open, they can be reused for shopping and to-do lists. By placing a magnet on the flap of greeting card envelopes, it makes a handy pocket to stockpile coupons on the fridge. When we plan the shopping list, the pocket gets a search to find any coupons that we could use.

Next time you need envelopes, stop in at the local card shop. Card manufacturers usually only accept returned cards from retail outlets, not the envelopes that come with them. These retail stores often sell these envelopes for a very low price. Because they are loose, there is no packaging to be thrown in the landfill and again you will save a little cash.

IN THE OFFICE

Many of us diligently recycle our waste office paper and should be proud of our efforts. For every ton of office paper that we take to the depot, about 380 gallons of petroleum is saved. There are many ways to reduce consumption and reuse office paper before it even sees the recycling bin.

In our home office, we keep a big box of paper that has been printed on one side only, for scrap paper. As writers, we use a lot of scrap paper. The box holds used computer, schoolwork, business flyers and any other kinds of one-sided paper. This paper comes in especially handy when children are visiting, keeping them busy with drawing and coloring. We use it for making lists, writing notes, or taking telephone messages. Lillian's family, having done this for generations, has come up with some nice ideas for the boxes in which the paper is stored. Her mother Joanne fashions a note-box made from waxed paper milk cartons, which when cut to size, is decorated with a spare piece of fabric and ribbon. For a child's room, use the comic section from a newspaper for wrapping the note box. Alternatively, wrapping paper and floral papers work just as well.

Cut letter size paper into quarters to use for smaller notes. We have a little antique wooden box that we use to store the loose paper by the telephone. In other areas, however, we choose to store them in a small enough box that can be tucked into a drawer—or set on a shelf with minimal clutter.

These same small pieces of paper can be made into a note pad by stapling a stack together. At 2 to 5 dollars for a new 100-page pad, this can easily save lots of money. Dave's dad, Frank, the ever-inventive crafter, has come up with yet another way to make notepads, which we just fell in love with. We asked him to share his idea with our readers:

"I cut waste printer paper into 4's, spread the 2 top and bottom ends (being the straightest) with *Modge Podge* glue. I found it works best if you put the pad on a flat surface and let it overhang a bit, so it is easier to put on the glue. Hold the pad down with any flat object to compact the paper and apply the glue. Modge Podge is great glue and it dries quickly." Frank also uses the glue to stick a small magnet to the back to affix the pad to a refrigerator.

Other businesses have started saving their one-sided paper as test paper for printers and found they could save thousands of dollars. Companies have also struck on the idea of printing their bills double-sided. To ensure their employees followed this policy, they set all the office machines to print double-sided automatically. This saved them quite a bit of money in paper, handling, storage, and postal costs. Print draft and informal work on one-sided paper. Even casual letters could be printed on reused paper.

Now, after both sides have been printed upon, office paper can be reused again as fire-starting material before being recycled. The paper can be shredded into 1-inch strips and reused for various things, but first, make sure the ink on them is non-toxic. Shredded, it can be added as a layer in the compost or as bedding for a worm bin. By using as an under layer in chicken coops and rabbit pens, or as lining for birdcages, the removal of accumulated waste will be made much easier.

These are very effective, money-saving methods for reusing waste paper that any home or office can easily carry out. Performing these few simple steps extends the life of the paper products before either recycling or converting them into humus by composting. It makes full use of the financial value of each page. Encouraging email communication reduces the energy of both the printer use and paper consumption. By using a fax modem (modems allow your computer to receive a fax without having to print it), you can reduce your paper waste and energy use even further.

JUNK MAIL

62 million trees and 28 billion gallons of water were used to produce US mail in one year, 50% of which is never opened.

While many of us diligently reuse and recycle paper, we are bombarded with junk mail. In 1990 alone, advertisers sent 62.8 billion pieces of unsolicited mail, including 12 billion catalogues, to Americans without permission or request. This equates to 34 lbs. of junk mail, or 1 tree, per person. It has been said that junk mail is as inevitable as taxes. It might seem that way, but we can get out from under this pile if we choose to.

There are times when we have received multiple copies of the same catalogue and other mailings from certain companies. This duplication is really an outrage; it is bad enough to receive one! However, these companies are very grateful to hear from you, as the duplication is only depleting their advertising budget needlessly. It only takes a few moments to write a letter, call, send a fax or email, yet doing so reduces the impact on our forests.

Fight the onslaught by mailing letters to the advertisers themselves, letting them know they are upsetting you as they cost you time and money to recycle the unsolicited mailings. Some individuals have gone as far as mailing the material back—enmass! That's a statement for sure—albeit an expensive one. First class mail may be returned by writing, *refused—return to sender* on the envelope. However, the post office will not return bulk mail.

Also, be sure to cancel any mail lists you may be on. You may not know you are on them, but if you have filled out a questionnaire, there is a good chance that company sold your name to a mailing list. Forms such as warranty cards and credit card applications, those attained when ordering products or services and entering contests, should all get special attention. If you do fill out any forms, be sure to check the tiny box (likely found at the bottom of the form in very small print) that states you do not want your name made available. Credit card companies, magazines, and manufacturers are often the most guilty for frequently sharing members information with other companies. Write them and ask them to refrain from passing on your name.

If former residents of your home are still receiving junk mail for you to deal with, you can do something about it. Fill out a change of address form at the postal office for each person no longer residing at the residence. When you fill out the change of address section, write something like, *moved, left no forwarding address*. You also need to state that the form has been sent by the current resident of the home (you) as an agent of the person named. Then sign your name.

Our research revealed some direct marketing companies that we have listed below. Whether you know it or not, you have probably received mailings from a direct marketing company at some time. Contact each of these companies and request to be taken off their mailing lists. The Direct Marketing Association estimates that being removed from their list alone can reduce mailings to you by up to 75% because of the volume of clients they have.

Whether you are contacting the companies on our list, catalogues, or other junk mail sources, you will need to know a few pointers that will speed the process. Often, there are codes corresponding with your name on the address label that will aid the service department in your request. Including the bar code and labels, or a photocopy of them along with your written or faxed requests will make them more effective. Some companies have toll free lines. Should you take advantage of these, you will hear a recording offering you several service options. Simply choose the one that will remove your name from their list. If they do not offer that option, chose the one that allows you to speak with a human. Again, have the bar code and label from the junk mail handy. You may receive a letter notifying you of the change or asking you to confirm your request.

Do not expect measurable immediate results. Be prepared to wait quite some time, as it can take from 6 months to 2 years to get off some of these lists. Be sure to keep their addresses handy, so that you can pass them on to others.

ADVO Inc.
Phone: #1-860-285-6100 Mail: List Services 1 Univac Lane Windsor, CT 06095-0755

American Family Sweepstakes
Phone: #1-800-237-2400

American Online
 Phone: #1-800-827-6364 (6am-2am EST)

Canadian Direct Marketing Association
Mail Preference Services, Concord Gate Suite 607, Don Mills, ON M36 3N6

Carol Wright
Phone: #1-800-67-target

Direct Marketing Association
Mail Preference Services, PO Box 9008, Farming Dale, NY 11735

Dun and Bradstreet
Phone: #1-800-333-0505

Metro Mail
Corporation List Maintenance, C/0 Customer Service, 901 West Bond Lincoln, NE 68521-3694

Publishers Clearinghouse

Phone: #1-800-645-9242 (8:30am-8:30pm EST) Fax: #1-800-453-0272 Address: 101 Channel Drive, Port Washington, NY 11050

Val Pac Coupons
Phone: #1-800-676-6878 Fax request to: #1-508-626-9925 Mail request to both: 1661 Worcester Rd, Framingham, MA 01701 and 1840 Aerojet Way, N. Las Vegas, NV 89030

MAGAZINES AND BOOKS

When you consider how many magazine publications exist in North America alone, the shocking use of our forests can really hit home. One ton of coated, high-quality, virgin magazine paper uses 15.36 trees. On the other hand, a ton of lower-end virgin magazine paper uses only 7.68 trees. With this in mind, subscribers need to write the editors of their favorite magazines, asking them to switch to an environmentally sound paper. To do our part, we can make efforts in various ways to reuse books and magazines before they are to be recycled.

To extend the life of books and magazines, tape the binding and the edges of the covers. This, while repairing any present damages, also prevents future wear and tear. The results, being a nicer look, could get a better price or have better trade value at a bookstore. Regardless, you have just extended the life of the book, allowing for lots of reuse.

Over the years, we have been involved with exchange loops—where we pass magazines on to other people, who in turn pass them on, and so on. It is a great, cost-effective way to remain updated and informed while extending the life of the material before it goes to recycling. We also enjoy a great variety of subject matter that we may not have been exposed to otherwise. Subscription fees are deterrents for many families who order only one, if any, magazine. Imagine what a family could save in subscription fees by participating in these exchange loops.

Both magazines and books can be donated to, or purchased from, fund-raising events. Quite often a school, library, church, or nonprofit society will hold these types of fund-raisers. Trade at used bookstores or sell in a garage sale. We once gave a whole box of magazines to a woman at a garage sale—she was very excited about getting so many. In addition, thrift stores will resell them to raise money for a worthy cause.

Any place with a waiting room or common area would welcome reading materials. Hospitals, offices, shelters for women, or missions for the poor are places that will readily accept them. Laundering facilities, often overlooked, are great locations to leave magazines and books. Schools may have need of your old issues as well—simply call the office to find out. Do not forget the local recreation centers and clubs, who often have programs and waiting areas that could make use of literary works. Places of employment often have coffee and lunch areas for the staff, where reading material is valued.

To help visualize the effect these efforts of reuse have on the environment, we would like to walk you through the life of some books we once had. Lillian's friend, Hildegard, gave us some boxes of books that her son had purchased from a used bookstore years before. After we read them, we passed them on to Lillian's mother. Then the books were donated to various thrift stores who sold them to raise money for charities or hospitals. These books were read at least once before they were sold at the bookstore, then by Hildegard's family, then us, then Lillian's mom, then on to a used store yet again. They had yet to see a recycling bin and have enriched many lives—and perhaps many more to come.

There is one other idea for magazine reuse that we would like to share. Take a few magazines, roll them tightly into a log and use several strong rubber bands to hold securely in place. Wrap the roll with recycled wrapping paper or cloth and place inside tall women's dress boots to hold the legs in place. You will no longer have creases and floppy boot tops to deal with.

NEWSPAPER

Very few of us realize how newspapers are affecting our forests. Canada supplies most of the world's newsprint and we harvest 247,000 acres of forest to meet that demand. Wood product consumption is depleting the world's rainforests, and according to the World Wildlife Fund, this results in three wildlife species going extinct per day.

Today, it is common knowledge that newspapers can and should be recycled at the local recycling depot. Producing from recycled paper uses 30-35% less energy and generates 95% less air pollution than making paper from trees. Recycling is not the only way to handle the problem of paper waste. There are many ways to reuse paper first.

Placing a thick layer of newspapers on naked soil and covering with a 2" layer of mulch (gravel, sawdust, or shredded bark) will protect the soil from erosion and weeds, while maintaining its moisture.

Placing shredded newspaper on the bottom of animal cages, chicken coops and rabbit pens makes for easy removal of waste and helps neutralize the richness of manure. Then the whole works can go in the compost. Use as its own layer in the compost heap, no more than 1" thick. Alternatively, shredded newspaper makes excellent bedding for worms. For more information on worms, see *Vermiculture*.

Single sheets of newspapers can be made into plant pots. Simply dip a sheet in water and fold to desired width. Wrap the strip around three fingers to make a cylinder. Then fold the end over, using the fingertips as support. Place in a flat so that additional paper pots support the seams. They can be transplanted along with the seedling when the time comes as it easily breaks down once planted. We have seen little paper pot molds sold for this purpose in seed catalogues, but for us, our fingers work just fine.

It was suggested to us that using sheets of newspaper to clean glass in place of paper towels worked well. We have found that this will result in some streaking, caused by the ink. For very dirty windows (wood stove door), one could wash with newspapers first, taking the worst of the dirt off and reducing the number of paper towels consumed. Crumpled newspaper sheets make an excellent cushioning material for delicate items when packing. However, here again, they will leave ink residue on the items so they will have to be washed upon unpacking.

We once had a wood stove, and during the winter used more newspaper as a fire starter than our mailbox received. Ask people that you know with wood stoves if they are in need of any of your surplus newspaper. Chances are they will be happy to take it.

When exploring the wilderness a ready source of dry material for a fire is not always available. Dave keeps several sheets of paper rolled up in a waxed paper cup with matches for this purpose. The waxed cup keeps the kit dry and is a super fire starter on its own – rip it into pieces and use to supplement the paper to start the fire.

Newspapers make excellent waterproof camping logs. To make a *newslog*, you need candle wax or recycled crayons, string, and several newspapers. Fold the single sheets so that they are no more than 8" wide. Roll one sheet at a time until the desired thickness is reached. The thicker and more tightly rolled the log is, the longer it will burn. Wrap the log tightly with a string made from any natural fiber. Place the wax in a metal can and set inside a second pot that has about 2" of water in it. Melt on the stove on a low setting. Coat the entire log with melted wax and when completely cooled, store in a plastic bag. Newslogs are lightweight and pack well in a backpack. They can be made any size and therefore are excellent additions to your emergency kit. At all times, we keep a kit in our truck and one in our backpack with the previously mentioned fire starter kit. You will be thankful that you prepared if any breakdown or emergency comes your way, and there is no wood source for a warm fire.

Use the comic section of your newspapers for a unique and attractive wrapping for gifts. If you put the wrapping on the box permanently and simply tie it shut with a ribbon, you will be encouraging the recipient to reuse the same gift-box again. For those who find the task of wrapping or unwrapping difficult, this box is a gift in itself. For more on wrapping ideas, see *Holidays*.

It is no secret that both telephone books and newsprint make great material for paper mache projects. A quick browse in the craft section of the local library will give you lots of ideas that children and families love for making other crafts like airplanes, hats, and boats.

PAPER BAGS

From large grocery bags to small wine and gift paper bags, you can reuse these in the most obvious of ways—as a bag. In our home, we use them to separate the paper recycling and to send guests home with fresh garden produce or baked treats. Flour and large pet food bags can be used as bags for sorting recycling. One bag for paper, one for cans, one for plastics, that kind of thing. This means that less time is spent sorting at the depot.

During seed harvesting season in our garden, these bags are employed liberally. Being paper, they breathe, allowing the seed to dry completely under protection from dust and insects. To ripen peaches, pears, plums, and nectarines, place the fruit in a bag, close loosely, and store at room temperature. Check every day for ripeness by pressing the fruit lightly with a finger. If it is soft and fragrant, the fruit is ripe.

Place store bought baked goods in a dampened paper bag in the oven to attain a fresh-baked appeal. However, some paper bags contain inks that may emit toxins during this process. Avoid any printed matter for food handling purposes. To reduce the calories in our home baked cookies, Lillian reuses the paper as an absorbent. First, she cuts them open along the seams so they lie flat. Then, placed so that the inside is up on top of a cooling rack, they are ready for use. The hot cookies are placed directly on the paper, allowing them to cool as usual. The paper absorbs the excess fat, which will be evident by the circular grease spots. This paper is not recyclable because of this, but it makes an excellent campfire starter. It can also be shredded and contributed to either the compost or worm bins.

Cut open the same way, and use the paper as parcel wrap—without the cost of the rolls available in stores. Scholars can reuse paper bags in much the same way—to protect schoolbooks, especially handy for the artistically minded for sketching on. Also when doing crafts, artwork,or any other messy chore, paper bags as a drop cloth will help with the chore of cleaning up. Depending on what the paper absorbed, it may still be composted.

Paper bags are valued as a material for children's crafts. Parents tell us they are a favorite for constructing puppets. Somewhat rigid yet workable, the paper is easily shaped, glued to, and painted on. If there are no children in your family, daycare centers and kindergarten classes are often on the lookout for such materials. A quick search on the Internet turned up lots of information on crafts using paper bags. Listed below are a few of the choices we found.

Paper Bag Craft Sites
- www.familyfun.go.com
- www.dltk-kids.com/type/paper_bag.htm
- www.rozani.com/mache.htm
- www.familycrafts.about.com/library/trcrafts/blpbag

PAPER TUBES

The production of paper tubes accounts for only about 7% of all paper produced, and favorably, includes a large percentage of recycled paper in its make up. Like all paper products, tubes are recyclable. Because of their handy shape, there are unique ways they can be utilized. Tubes come from products such as foil, plastic and waxed wrap, paper towels, wrapping paper, and even printer rolls and ribbons.

Toilet paper tubes can be stuffed with a set of stockings and kept in the glove compartment or office desk for emergencies. Use them for wrapping small gifts—insert the gift and decorate with wrapping paper. They can be donated to day-cares and preschool or kindergarten classes for craft projects. The small tubes make excellent pet toys as well. Our cats love to chase a paper tube as it tumbles down the stairs. Paper towel tubes keep hands at a safe distance when playing with overzealous felines who love to grab at and kick the other end. Hamsters, apparently, love to chew and shred them.

Create sidewalk chalk for children using small tubes. Begin by taping wax paper over one end to block the hole. Stand the tube upright on the wax paper side, pour plaster of Paris into the tube, and allow to dry completely before handing over to the kids. The paper roll is left on keeping hands clean, and as the chalk is used, simply peel back more of the roll.

For gardening, the thin walled tubes make good biodegradable seedling pots. Cut to size, pinch one end closed, and place in a seedling flat. Fill with soil and plant your seeds. Carefully work a section of tube over a new transplant for a sturdy cutworm collar that falls apart in about 4-6 weeks. By then, the plant is tough enough to be impervious to the cutworms. The tube can then be added to the compost pile. Use gift-wrap tubes (longer and larger in diameter) to slip over leeks and celery plants in the garden. By forcing and twisting with your hands, screw the tube into the soil about 1". The tube will keep the leek stem shaded and replace the traditional blanching method of continually piling soil against the plants growing stem. The tube reaches further up the plant than the traditional methods, so the leek grows a longer white section and has fewer pests.

Paper towel rolls make excellent storage containers for plastic shopping bags. They can hold about 8 bags, and this can come in very handy when there is the need for a small garbage bag. We keep one of these bag-tubes in our vehicle and one in the backpack. This way there is no excuse for us to pass that garbage by as we hike along our favorite trail, as in *Clean Walking*.

Sturdy paper tubes work well to store your recycled wrapping paper. To remove creases and flatten recycled wrapping paper, begin by laying it face down on a safe, flat surface. Lay a towel on top as a buffer to prevent burning. Then, with an iron on the lowest setting, steam the recycled wrapping paper to remove creases. Then roll it onto a paper towel or old gift-wrap tube. Close by winding a ribbon around the paper, or by placing elastic bands over it. Lillian's mom stuffs ribbon inside the tube itself. She calls it her wrapping kit because the tube holds all the materials she needs.

Several sizes of tubes work well as sleeves to store extension cords and wires. Wind your cord into a coil and bend so that it feeds into the appropriately sized tube. Slide the tube so that it rests near the center of the cord. We have found that toilet paper tubes are just the right size for small electrical and stereo wires. Larger tubes work best for long extension cords.

Ever have difficulty cleaning under or beside your appliances, cupboards, and furniture? Well, we have developed a way to ease this chore. A 1 1/4" wrapping paper tube typically fits vacuum hoses nicely. Flatten the opposite end gradually, but not so flat that items cannot pass through. It is a handy tool for getting into those narrow, hard to reach areas.

Pringles potato chip containers are very handy for travel and outdoor activities. With a plastic lid, they are perfect for keeping crackers, biscuits, or cookies intact in a crowded food hamper or backpack. Lined with a recycled plastic bag, they can be used to store bulk kitchen items like pasta noodles or rice. In the workshop, they can store rags, bits of string, and loose parts.

Because they are lightweight and absorb the shock of rough handling during transportation, paper tubes make an excellent packing material. The recipient can then recycle or reuse the packaging for another parcel.

PAPERBOARDS

With paperboards accounting for 50% of all paper products, isn't it time we found ways to extend the life of this product, reduce the consumption, and recycle every little scrap? Our recycling efforts do make a major dent in our energy use; for every ton of paper products that we recycle in Canada alone, we save 4,100 kilowatts of energy. Boxes, fortunately, are one of the easiest items to find reuse ideas for before having to be recycled.

Corrugated cardboard boxes are obviously first choice for shipping out a parcel or wrapping a gift. Often when moving, these big strong boxes are hard to find and sometimes even have to be purchased. It would be nice if stores advertised that they have good sturdy boxes they are recycling. It could save them the time it takes to flatten the boxes and take to the bin, while possibly providing an essential service for a minimal cost.

Large, sturdy boxes make excellent pet beds when lined with a snuggly old blanket. Dogs and cats are not the only mammals that enjoy reusing boxes though. Most parents know that young children find big boxes fascinating, sometimes more so than the gift itself. Many of our friends have taken the time to build dollhouses and forts for their children out of large appliance boxes. This is a great way to spend quality time with the kids, while saving money on toys.

At our rural home, we have brush and weeds in our garden pathways that we are constantly battling. A double layer of heavy cardboard boxes helped us deal with this. First we laid the boxes out flat and removed any tape or staples. Then we watered them down to make them pliable and able to form fit to the earth. Lastly, we covered the area with a thick layer of bark mulch. Because cardboard will break down in about a year, the weeds will smother while the soil becomes enriched with organic matter. Meanwhile, the bark mulch is very pleasant to look at and walk on.

Corrugated cardboard cut into squares makes a good shock absorber when used for boxing up jars, glasses, or vases. Layer the squares between the glasses—preventing cracks, chips, or breakage during moving or storage. An artistic gardener friend cuts out cat shapes on cardboard panels, paints a cat image on either side, and places them throughout her fruit trees. She claims it is an effective bird deterrent, but admits that it does not endure long bouts of moist weather. Nevertheless, she enjoys creating a fresh herd of cats each year.

Making educational puzzles from single-ply paperboard boxes is a good fun project to do with your children. Glue a map onto a flat piece of paperboard. When it is completely dry, turn the puzzle over and draw the design of the puzzle pieces to be cut. This chore is easiest with a razor knife, so be sure to cut on a scrap board or something that can be scratched up by the knife. Piecing together the puzzle also teaches the child familiarity with the map. Another craft for younger children is to wrap small pudding or *Jell-O* boxes in recycled wrapping paper and use as stacking blocks. Tissue boxes can be cut and used for cassette and video tape storage.

Large, sturdy cereal boxes and laundry soapboxes can also be reused as magazine holders. Cut off the top flaps and on one end panel measure up 4 inches from the bottom and draw a horizontal line. On the top of the open end measure 2 inches across and make a mark. Connect with a diagonal line the top mark to the bottom mark. Repeat for the other side. Cut along these lines to make the magazine holder, sending the small pieces to the recycling bin. This can be covered in material (recycled denim) or in wrapping paper. It will be necessary to glue, staple, or tape the covering in place. Any seams can be hidden with some ribbon and a glue gun.

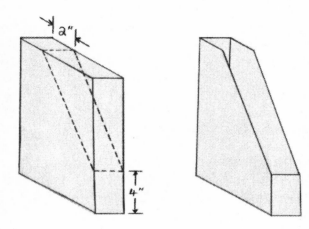

Boxes can also be made into permanently wrapped gift boxes, which can be reused many times. See *Holidays*. When shredded into 1/2" strips, single ply or corrugated paperboard acts as excellent bedding for a worm bin or as layers in the compost.

Let us not forget the paperboard egg cartons. If you can not find a farmer to take them, recycling should be next on your list of options. However, if you are an outdoor enthusiast or travel by road often, then you may consider making an *egglog*. You will need a paper egg carton, 12 briquettes, dryer lint or cotton balls, and some candle or crayon wax. On a low setting, melt the wax using a metal can in a pot of water as a double boiler. Fill each egg indentation half full with melted wax. Press a briquette, lint, or cotton in each indentation—we have read that some people dip the cotton or lint in petroleum jelly, but do not practice this ourselves. When cooled, close the lid and store until use. Break off individual eggs for a fire starter or use the whole log for a longer burn time. In wet conditions, remove the lid and use it as bedding for the egglog to sit upon.

Extremely light, this log can be stored in a recycled plastic bag (ensuring dryness) in the backpack or in the car as part of your emergency kit.

PART
FOUR

THERE IS HOPE

As a final note to this book, we would like to point out that although it is a big task to change the thinking of such a mass of population, there is hope. The collective actions of many people are much more powerful than that of any one high-ranking politician. As the majority, we can force changes by making changes. If one can weed through all the bad news that is spewed at us on a daily basis and experience some good news, it can be quite refreshing— inspirational even. Therefore, we leave you with some inspiration to start right where you are.

Recycling has really caught on and already we can measure the results of our combined efforts. Recycling paper reduces the paper industry's energy use by 35% and its air pollution by 95%. For every ton of recycled paper, 17 trees and 3.3 cubic yards of landfill space are saved.

National Polymers Inc. figured out the impact a 100-unit apartment building would have on the environment if there was maximum recycling participation over one year. They would save 21.93 thirty-foot trees, 8,389 kilowatts of electricity, 26.86 cubic yards of land fill space, and 77.4 lbs. of air pollution.

Canada is considered the world leader in protecting the ozone layer. The use of ozone-depleting chemicals has been reduced by 95%. The economic savings of Health Care Benefits for Canadians are estimated to reach as high as $10 billion annually, *if* the Kyoto Agreement is met.

The city of Toronto proved that this is attainable. They have cut emissions by 67% since 1990, exceeding their original goal by 3 times—one of the largest achievements in Canada. Simply by improving the energy efficiency of city buildings, they cut energy use by 8%. The Green Communities Association of Ontario asserts that 25% of energy reduction would result in 4.4 tons of Green House Gas reduction. Part of the solution will come from the new energy efficient buildings available today that save up to 90% in energy use. Toronto also replaced all city lighting with high efficiency bulbs. Their biggest reduction came from installing gas collection pipes in the landfills where millions of tons of garbage emit Methane—considered 21 times more environmentally harmful than Carbon Dioxide. The pipes transport the gas to generating stations that provide enough power for 7,000 homes.

Calgary Transit's *Ride the Wind* program was the first public transit system to be fully powered by wind. It has been so successful that use has increased 73% since its opening in 2001.

The Canadian Wildlife Federation has helped seniors groups create, enhance, and conserve wildlife habitats. Their *Wild About Gardening* program has helped thousands of backyard gardeners apply their skills in this way. CWF's *Communities for Wildlife* program also helps get the community involved in conservation projects.

Many public schools have successfully raised funds to convert areas into greenery by selling T-shirts, quilts, and crafts volunteers have made. These areas become environmental learning centers where wildflower patches, along with Aviary areas, are part of the curriculum. The children learn about birds, insects and botany, as well as how to use the plants for cooking, herbal oils, vinegar, and dyes.

In 1991, *Trees For Life Canada* (a charitable organization dedicated to planting trees) developed the *Grow-a-tree* program for elementary students in hopes of them attaining a respect for the environment. Over 23,000 trees were planted in 1993 alone thanks to this program, which in 1995 grew to include 53 schools in China.

In Ottawa, where they have the *Take it Back* program, people are connected with places that accept old automotive parts, electronics, garden supplies, health products, and household products. In 1999, Holland created a law requiring that appliances, computers, and other equipment be returned to the manufacturer.

Packaging waste accounted for about 1/2 the volume in Germany's landfill, until they shifted the responsibility to the manufacturer. Germany's national landfill contribution has now been reduced by 60%.

Manufacturers are learning that recovering waste reduces operating costs. Among them is the Canadian Battery Association who developed, in cooperation with the Rechargeable Battery Association, a battery recycling program called *Charge Up to Recycle*.

Epson recycles 90% of its waste, disposing of the rest at a waste-to-energy facility.

Fetzer Vineyards reduced waste by 93% by, among other things, composting corks and organic waste.

Mad River Brewery diverts 98% of its waste from the landfill and takes back its containers.

BellSouth Telecommunications are reducing their paper use by 1.3 million pounds simply by printing its customers' bills double-sided.

Xerox has reached an 88% recycling rate, saving the company millions of dollars.

Pillsbury reduced their waste by 96%.

RESOURCES

This book has taken many years of research, experimentation, and interviewing individuals. Our resource list expanded to a huge number of articles and studies from a vast selection of magazines, newsletters, web sites, and pamphlets. Instead of tackling the terribly involved task of listing each and every article and study title with footnotes, we are simply listing the sources here.

We urge the readers to use the list provided to continue their own journey. Be prepared that some of the web site resources may no longer be accessible due to the fast paced changes in the world of the Internet.

Green Adviser <www.greenadviser.org>

<www.recyclemore.org>

'The Home Owner's Guide to Planting Energy Conservation Trees' by Friends of the Earth
 <www.epa.gov.epaoswer/non-hw/muncpl/reduce>

<www.reduce-reuse-recycle.com>

<www.ecoaction.ca>

<www.climatechangesolutions.com>

<www.recycle.nrcan.gc.ca>

Alliance of Foam Packaging Recyclers <www.epspackaging.org>

The Preserve Toothbrush Subscription <www.recycline.com>

Grass Roots Recycling Program <www.grrn.org>

Planet Save <www.planetsave.com>

Waste-line <www.waste-line.flora.org>

<www.geocities.com/RainForest/5002/index.html>

<www.epa.gov/students/municipal_solid_waste_factbook.htm>

<www.adpc.purdue.edu/PhysFac/why.htm>

<www.mcswmd.org/kids>

<www.dist428.dekalb.k12.il.us/lincoln/club/recycle.htm>

<www.recycle.ubc.ca>

Environmental Systems of America <http://envirosystemsinc.com/resrec.html>

Purdue University <www.adpc.purdue.edu/PhysFac/why.htm>

<www.interiordec.about.com/library/weekly/aa042201a/htm>

2 Minutes A Day For A Greener Planet by Majorie Lamb (1990 Harper Collins)

<www.encorpinc.com>

David Suzuki Foundation <www.davidsuzuki.com>

The Green Communities Association of Ontario

CBC Radio

Earth times news Service <www.earthtimes.org> on-line newsletter

Eren <www.eren.doe.gov/consumerinfo/energy_savers>

Packaging Council <www.polystyrene.org>

<www.grn.com/chat/talk.htm>

Environmental Technology Center <www.etcentre.org>

<www.ec.gc.ca/science>

Earth Island Org. <www.earthisland.org/bw/25_percent.shtml>

EnerGuide <oee.nrcan.gc.ca/Equipment/english/index.cfm?PrintView=N&Text=N>

Blue Water Network <www.bluewaternetwork.org>

Natural Resources Canada Office of Energy Efficiency <http://oee.nrcan.gc.ca>

Sierra Club <www.sierraclub.ca>

How Stuff Works <www.howstuffworks.com/two-stroke3.htm>

<www.recyclingtoday.com>

Recycling Inc. Report: *Waste @ work: Prevention Stragaties for the Bottom Line*

Divorce your Car by Katie Alvord

Biocycle - Journal of Composting and recycling: <www.jgpress.com>

Car Busters: <www.carbusters.ecn.cz>

Highway Users Alliance Report: *Unclogging America's Arteries: Prescriptions for Healthier Highways*

Culture Change <www.culturechange.org>

Health Magazine

World Wide Fund for Nature

Breaking Gridlock: Moving Toward Transportation that Works by Jim Motavalli

The Globe and Mail

Dr. Jane Goodall's E-newsletter

Economic Policy Institute & Center for a Sustainable Economy Report: *Clean Energy & Jobs: A Comprehensive Approach to Climate Change & Energy Policy* <www.sustainableeconomy.org/press/globalwarming8.pdf>

David Suzuki Foundation & University of Victoria Report: *Up in The Air* <www.davidsuzuki.org/files/Ozone.pdf>

The Humane Society <www.hsus.org>

Organic Gardening Magazine

<www.globalstewards.org/main.htm>

Gardens West Magazine

<www.marinwater.org/swimmingpools>

<www.container-recycling.org >

<www.bottlebill.org >

<www.ecocylce.org>

<www.nrc-recycle.org>

<www.zerowasteamerica.org>

Environmental Magazine: <www.emagazine.com>

www.recyclecongress.org>

<www.BAN.org>

<www.ZERI.org>

<www.Rachel.org>

<www.WorkOnWaste.org>

<www.oldgrowth.org/compost/forum-vermi/>

Mary Appelhof <www.wormwoman.com>

<www.mastercomposter.com>

Halifax Regional Municipality, Waste Resource Division <www.region.halifax.ns.ca>

Take it Back Program, City of Ottawa <www.city.ottawa.on.ca/ gc/takeitback/index_en.shtml>

Citizens' Network on Waste Management <www.web.net/ cnwm>

Toronto Environmental Alliance <www.torontoenvironmet.org>

Ministry of the Environment, Waste Management policy Branch <www.ene.gov.on.ca>

New American Dream <www.newdream.org>

The Country Connection Magazine <www.pinecone.on.ca>

<www.treegivers.com>

<www.arborday.org> National Arbor Day Foundation

<www.treesftf.org>

<www.greenlinepaper.com>

<www.ecopaperaction.com>

Conservatree 100 2ⁿᵈ Avenue San Francisco, CA 94118 Email: <paper@conservatree.com> Web: <www.conservatree.com>

Residential Recycling Factoids <http://envirosytemsinc.com/

resrec.html>

Recycle City <www.epa.gov/recyclecity>

<www.containerrecycling.org>

Earth Ship Homes <www.earthship.org>

Aquila Networks Canada - Powerlines Newsletter <www.aquilanetworks.ca>

<www.earth911.org/master>

Regional District of Kootenay Boundary <www.rdkb.com>

<www.usedoilrecycling.com>

Backpacker Magazine

<www.raincons.org>

<www.livos.com>

<www.polywood.com>

<www.trex.com>

<www.usplasticlumber.com>

Countryside Magazine

Worm interview by Mike Rotina <http//harmonycentral.org/data/text/channels/soilfolder/wormwoman>

American Forests <www.amfor.org>

Us Forest Services' Urban Forest Ecosystem Research Unit <www.fs.fed.us/ne/syrcuse>

Environmental Defense <www.environmentaldefense.org>

<www.simpleliving.net>

<www.greenmatters.com>

<www.use-less-stuff.com>

<www.rdn.bc.ca>

Trees For Life Canada - 143 Cayuga Ave Ancaster, On L9G 3B2

Association of British Colombian Professional Foresters Email: <dyochim@rpf.bc.org> Phone: (604-687-8027)

Rays Foods <www.abundanthealth.info/xotherproducts/disaster-preparedness>

Territorial Seed Company PO Box 158 Cottage Grove, OR 97424-0061 <wwwlterritorialseed.com>

Seeds of Diversity Canada Box 36 Stn. Q Toronto ON M4t 2L7 <www.seeds.ca>

Trees For Today & Tomorrow - 44 Eglinton Ave. West Suite 206, Toronto, ON M4R 1A1

The Evergreen Foundation - 24 Mercer StreetToronto, ON. M5V 1H3

<www.epa.gov/epaoswer/non-hw/recycle/jtr/comm/glass.htm>

Washington DC – Container Recycling Institute

<www.epa.gov/grtlakes/seahome/housewaste/src/glass.htm>

Inform and the Council on the Environment of New York

<www.globalstewards.org/junkmail.htm>

The Energy Efficiency And Renewable Energy Network

<www.retiring@lgov.com>

<www.interiordec.about.com>

<www.redjellyfish.com>

<www.wildaboutgardening.org> Canadian Wildlife Federation's site on backyard gardening.

<www.cwf-fcf.org> Canadian Wildlife Federation Online Newsletter

Trees Ontario - 150 Consumers Road Suite 502 Willowdale, ON. M2J 1P2

<www.gettoknow.ca>

<www.jadeandpearl.com>

<www.organicessentials.com>

Lunapads International 207 West 6th Avenue Vancouver, VCV571K7 <www.lunapads.com>

Printed in the United States
20318LVS00001B/476